Health, Culture and Society

Health, Culture and Society

A Comparative Analysis of French, Chinese and Indian Society

Bernard JOUANJEAN

Rawat Publications

Jaipur • New Delhi • Bangalore • Hyderabad • Guwahati

ISBN 978-81-316-0397-0

First published in 2009 in French under the title *Physiologie du risque face à l'Histoire*.

Translated from French by Cressida KEHER.

Published by
Prem Rawat for **Rawat Publications**
Satyam Apts, Sector 3, Jawahar Nagar, Jaipur 302 004 (India)
Phone: 0141 265 1748 / 7006 Fax: 0141 265 1748
E-mail: info@rawatbooks.com
Website: www.rawatbooks.com

New Delhi Office
4858/24, Ansari Road, Daryaganj, New Delhi 110 002
Phone: 011 2326 3290

Also at *Bangalore, Hyderabad* and *Guwahati*

Typeset by Rawat Computers, Jaipur
Printed at Chaman Enterprises, New Delhi

Contents

Acknowledgements

I would like to thank

Mr Claude Hagège, Professor of Linguistic Theory, Collège de France, for having taken every dimension of the word 'prevention'.

Madame Anne Szulmajster, Celnikier, Engineer, 4th section, *l'École pratique des hautes études (EPHE)* and Chairperson of Linguistic Theory, Collège de France, for her attentive listening and wise advice.

Miss Madeleine Biardeau, Director of Studies at the Fifth Section of the EPHE, for having shared her vision of India with me.

Professor Kanhaiya Lal Sharma, Vice Chancellor, Jaipur National University, Jaipur, for his attentive and kind support.

Preface

It is not completely coincidental that the author of the book you are about to read is, notably, the physician of the Collège de France. That is to say of an institution where the laboratory of the renowned expert Claude Bernard was located. In fact, the following pages recall with authority, and not without courage considering the inertia which they call into question, some of the key ideas of Claude Bernard in his major work of 1865, *Introduction to Study of Experimental Medicine*. What the American physiologist Walter Cannon was to call homeostasis in 1929, was none other than the stability of the internal environment highlighted in the great book of Claude Bernard.

However, on these solid premises, Bernard Jouanjean offers an original construction which, in many respects, goes still further. Because it is at the base of an essential and still largely ignored concept (both in medical institutions and in practice), namely, the concept of prevention, he asserts that of regulation, or balance between the various biological factors, which is carefully and indefinitely maintained. He shows us, with clarity and insistence, that all medical researches, whether explicitly stated or only implied, have no other purpose than to develop and even to exalt, whenever possible, the forms and instruments of balance, which belong to the natural state of the human body.

Suffice it to say that what we have here is a fundamental book. In fact, the author sees prevention, the meticulous preservation of physiological balance, as the basic therapeutic principle and not simply as an auxiliary action of therapy. Thus, we see here, the clearly emerging, social and ethical dimensions of medical activity, completely articulated through the preservation of the well-being of human societies. Certainly, one cannot imagine that doctors will not be convinced by these brilliant ideas, which exhibit the talent of the author and the strength of his conviction and which convey the image of the "simple" truth. But until now, no one, even among those who are willing to recognise the importance of these ideas, has gone to the extent of the author, who devotes an entire book. Just as much as doctors, the general public must also feel very concerned about such an essential problem as that of public health. The conclusions of the author have perfect logic and contain strongly innovative ideas, to say the least: prevention is closely linked to the curative act itself, since it has to anticipate pathological damage by a careful and thorough study of risk factors. Moreover, the notion of health is no longer separable from that of prevention.

However, in addition, the approach of this book is characterised by another very particular originality. Not content to emphasise the considerable importance of this notion of stability or homeostasis purely from the physiological and medical point of view, he highlights its deep connections with social representations. Long interested in the culture of classical India and that of ancient China, he acutely reveals how the concept of socio-ecological balance is built in these two great eastern cultures. Developing an observation nourished by his keen curiosity of these areas of civilisation, of which others often have only a superficial knowledge, or are provoked by the exoticism of poor quality goods, the author wonders, in pages full of finesse and of profound insights, about the adequacy or the contradictions between bipartition or tripartition of social functions on the one hand, and the characteristics of these societies which he describes as low- or high-risk, on the other.

One might ask what relevance such considerations and historical analyses have to the central problem of prevention, to which the author, doctor of prevention, devotes his professional activity as well as his intellectual engagement. In fact, the relevance is obvious if one is willing to be convinced of the value of the arguments outlined above, which establish the logical and factual interdependence between prevention and maintaining balance.

Everything is built on a deep desire of Barnard Jouanjean, who defines his personality and expresses himself in words not devoid of beauty and of evocative power: the quest for harmony.

December 2010

Claude Hagège
Collège de France

1

Introduction

Our world currently suffers such turmoil that it is impossible not to worry about the future. However, some would tell you that such anguish has also been known in previous centuries. The scientific knowledge acquired during the 20th century has profoundly changed the scale of risk incurred by man and today the survival of our civilisation is at stake. Since the middle of the 20th century, we have entered into a civilisation of high risk for life itself, and we will suggest, in the development of this essay, a physiological definition.

The constant search to improve our living conditions gives rise to increasingly sophisticated achievements, thanks to technological progress, the effects of which on the body and the environment are often poorly understood. Within industrial complexes, usually highly polluting, man produces objects which make it possible to extend the expression of our physiological functions. The enthusiasm they generate and their distribution to the general public should not obscure the degree of ignorance about their long-term consequences.

Promoted by advertising through the various media, these "consumer goods" upset the foundations of our society a little more each day. By removing certain social cues, which were patiently constructed from generation to generation, they are likely to cause psychological imbalance, accompanied by behavioural disturbances, sometimes even psychiatric. In addition, industrial pollution and the use of certain inventions can cause physical damage resulting in allergies, cancer and damage to the immune system, etc. In becoming aware of all the dangers – current and potential – some people have reacted sharply and started understanding the need not only to respect but also make respected a natural balance, whether environmental or behavioural. In fact, our survival depends on it.

According to Alain Rey, "The word 'balance' has been used since the beginning of the 17th century (1611) for 'equal weight', then as a scientific term (1663, Pascal) to designate the fact, for several forces acting simultaneously on a material system, to not change its movement or at rest state, thus by metonymy[1], the state of such a system. By extension, the word commonly means 'stable attitude' with the adverbial phrase in balance. Balance is also commonly used (1674) in the field of the arts in the sense of harmonious organisation of composition. Since the early 18th century, the word indicates a relationship considered as good between opposite things."[2]

---

1. Metonymy: Language process with which we express a concept by means of a term indicating another concept, which is linked to it by a necessary relation (the cause for the effect, the container for the contents, the sign for the signified thing), *Le petit Robert*, 1995.
2. Alain Rey, *Dictionnaire historique de la langue française*, (Historical Dictionary of French Language), 1992.

By "studying the functions and properties of the organs and tissues of living organisms"[3] the physiology (gr. physis – nature; logos – speech) characterises a state of balance of the body which may be disrupted by risks.

What is particular in the analysis of physiologic risk is that man will suffer the effects of a situation for which he could be responsible. He would consequently be considered solely responsible for an event which is detrimental to his own balance. This awareness of risk and of its consequences may be based either on an assumption or probability, or on an actual experience, blending in with the memory of old actions which must not be reproduced.

In view of the growing power of risk, the word 'prevention', for a long time, discredited and often rejected because of the absence of immediate profitability, and has rather offered research themes in domains largely beyond the medical framework. Regulation and balance, which are imperceptibly associated with the word 'prevention', add a much broader scope of intervention to its definition which till now has not been admitted.

How could we fail to question our somewhat recent and sudden realisation of the fragility of human balance in its environment? How could we avoid aligning our vision of the future with the messages of prevention transmitted with an obsessive tenacity by generation after generation of Indian and Chinese societies? How can we fail to be astonished by the gravity of these messages which have hammered the memories of many people and modified their behaviour over centuries? What are the limits of the adaptation of the human body to environmental change?

---

3. Garnier-Delamarre, *Dictionnaire des termes de médecine,* (Dictionary of Medical Terms) 22nd edition, Maloine, 1989.

The obsession of balance reveals a behavioural paradox reflecting an opposition between rigidity and flexibility: on one side it expresses a rigorous attitude, establishing a typical framework which serves as a reference mark and as a wall of protection; on the other side it is defined as a principle of regulations relying on adaptation mechanisms. The fear of seeing an event happening can become an obsession if it is judged to have such gravity that it can take on the future of our humanity. Fear of the future and obsession of the past in extreme situations form two inseparable concepts.

In discussing the theme of the book, Jean Delumeau, Professor of the Collège de France, pointed out to me that this work treated the physiological balance of man according to history. Physiology is studied in healthy subjects and serves as reference to define a healthy condition, i.e., a state of balance in its entirety. Beyond a generic definition of the word 'balance', this approach integrates the idea that one cannot speak about balance without taking history into account. Indeed, it is possible to relate the physiology of risk with the development of human societies throughout their histories.

In order to consider the types of events which could generate an obsession with physiological balance, we will question instances of reactive behaviour that a cataclysmic event could cause, analyse the factors which conditioned the behaviour of ancient Indians and question whether an obsessive desire for balance can be spontaneous or reactive. These are the objectives which will be considered successively throughout this text. Within the framework of this reflection, we will introduce the concept of 'antithetical memory'.

The miracle of our stability, and consequently our identity, finds its source innately in our constitution. It is also due to our immune defence system, which has everything to determine our physical and psychological balance within our environment. Our actions, which modify our experience, come to confirm or invalidate our state of balance. Consequently, an essential question arises: how do we maintain this state of balance durably? Or, how do we connect physiological stability to social regulation?

2

Physiology, Regulation and Identity

Regulation and Stability

Self-recognition and finding one's way in a landscape constitutes in itself a proof of stability. This fact is more of an observation than an artificial construction. We define this situation as being very particular, and yet at the same time, as being as universal as a state of balance. Without stability, we would have no reference point. It would be a mess.

However, the stability of living beings does not present any static character. It is a result of a combination of autonomous, biological mechanisms of regulation and the voluntary acts likely to preserve or modify it. This state varies over time according to an invisible dynamic. To observe it, it is necessary to study its fluctuations and establish correlations according to the appearance of organic modifications. While any chemical reaction has a state of balance, which is either stable or unstable according to the presence of components lacking the capacity to reorganise, the cells of the organism are reconstituted according to a definite model. Throughout our entire life we have an identity, retaining the same face, even though all of our cells are continually being renewed.

A design of physiological regulation only appeared in France in the middle of the 19th century with the publication of Claude Bernard's[1] work *Introduction à l'étude de la medicine expérimentale* (Introduction to the Study of Experimental Medicine). He highlighted the biological stability of the organism by studying the glucoregulation of the liver and thus became aware of the existence of the laws which determine our identity. According to him, "Determinism, which means the identity of effect related to the identity of cause, is a scientific axiom which could not be violated more in life sciences than in sciences of inanimate objects (astronomy, physics and chemistry)". Thus, we are conditioned to be what we are by laws, and we can only work on ourselves by taking these principles into account. Bernard further says, "We can control the phenomena of nature only by submitting ourselves to the laws which govern them."[2]

However, in the body, nature combines a certain number of functions, some of which could be antagonistic.

---

1. Claude Bernard (1813–1878), medical doctor and physiologist, Professor at Collège de France, Sorbonne and National Museum of Natural History, is considered as the founder of experimental medicine. He highlights the "stability of the interior environment", state of balance that Walter Canon, an American physiologist of 20th century named "homeostasis".
2. Claude Bernard, *Introduction à l'étude de la medecine expérimentale* (Introduction to the Study of Experimental Medicine), Champs Flammarion, 1984, p. 128.

Louis Pasteur[3] thus proved the defensive action of the immune system against a foreign antigen. The pathogenic effects of the rabies virus on the body could, however, be suppressed by vaccination, characterising at same time the existence of an autonomous defence system and also of an immune memory which protects us from external viral and microbial aggressions.

Thus, the peaceful coexistence of the immune system with other functional components within the body signals an implicit knowledge of "mine" and "not mine". This association of functions reinforces the idea that the identity is specifically defended at every moment for every individual. It expresses the resulting dynamic of the mechanisms of regulation concerned with homeostasis – the state essential for survival and reproduction. The preservation of this stable condition throughout life makes man an autonomous being.

It should be specified that Claude Bernard insisted particularly on the determinism of the laws and the phenomena accompanying them without mentioning the determinism of the results. Hence, the relevance of a scientific experiment, which always requires a hypothesis, must be compared with the experimental method and be verified. It is based on a theory and does not have the logic of systems built on logical reasoning, which have not been subjected to proof against experimental facts. There remains much uncertainty about the results of an experiment. Bernard writes: "The systems therefore tend to control the human spirit, and the only utility that can, in my opinion, find

3. Louis Pasteur (1822–1895), French biologist and chemist, is known in particular for his work on the vaccine for rabies. He provided the foundations of experimental methods in microbiology, highlighted the role of microbes in the propagation of disease and invented pasteurisation.

them, is to provoke fighting which destroys them in shaking up and exciting the energy of science."[4]

The objective of experimental science is more to adopt a method facilitating the observation of the effects of the laws and the interpretation of the results. This helps to better understand their purpose in the tangle of reactions, which establish, for example, a stable state. Admittedly, we only discover mechanisms in a scientific reductionist approach, with such complexity that the synthesis of the results on the body is difficult to understand. Thus, an anti-hypertension drug will be able to appear effective in vitro, and will present a set of side effects in the body representing the disturbance that it generates on a global level, hence, the interest to compare the results of an educational method with those recorded by a global or holistic method.

The stability of an individual, if it can be envisaged on a theoretical level, is completely dependent on the complexity of organic and environmental interactions. Because of the importance of variation, it limits any form of predictability and selectivity. However, it introduces the concept of capacity of action and functional predisposition.

Our body, plunged for millions of years in a variable ecosystem, has had to adapt to environmental change. It is what led Charles Darwin[5] to formulate a theory of evolution explaining, by the mechanism of natural selection, the laws of adaptation. However, this evolutionary theory, if it is accepted as a general principle, has flaws, highlighted by

4. Claude Bernard, op. cit., p. 309.

5. Charles Darwin (1809–1882), English naturalist, known for his work and theories on the evolution of living species.

Stephen Jay Gould,[6] who defends a theory of punctuated equilibrium. Moreover, and it is a personal metaphor, this theory does not consider a regressive principle, where, by a mechanism of regulation, the organism "would decide" to protect certain genes unsuitable for a sudden change of environment by storing them without expressing them. They would thus constitute reserved genes having memorised a specific environmental situation. This evolutionary choice with phenotypical transformation would correspond to a non-expression of certain genes and could support adaptation to a new environment. In any case, it would undoubtedly have physical and mental effects on the bodies concerned. The genes put in reserve would await an environmental change, more conducive to their expression, in respect of homeostatic balance. Thus, the body could change morphologically (phenotypic[7]) while preserving its former genetic potential, constituting a de facto "genetic memory". In this case, is it an evolution or a regression? Changes can either seem like regressions degrading physiological balance, or like functional improvements according to the increase in capacities of human physiological

---

6. Stephen Jay Gould (1941–2002), palaeontologist, contributed to the theory of evolution in criticizing the fundamental concepts of Darwinism by proposing in 1972 the theory of "punctuated equilibrium". It puts forward the assumption that the history of evolution concentrates in relatively fast episodes of speciation instead of occurring in the form of slow and continual transformations of established lines.
7. The phenotype is all of the observable characteristics (anatomical, morphological, molecular, physiological, ethologic) characterising a given living being (colour of eyes, hair ...). The phenotype is dependent on the genotype which constitutes the information carried by the genome of an organism (DNA) carried by the chromosomes inside the nucleus of cells. It is influenced by the environment and translates in the expression of the genes.

adaptation. One question is essential: as the body must implement the capacities of adaptation to a new medium (forced learning), can we speak about enriching the gene pool during a change affecting the physiological balance? Wouldn't there be a graduation between non-expression of genes and the appearance of irreversible changes? Wouldn't the human organism have experienced such transformations over centuries?

"The wisdom of the human body is the result of the wisdom of the genes."[8] Thus, following the example of Walter Canon who defined the wisdom of the body in a book called *The Wisdom of the Body,* Christopher Wills was interested in the mechanisms of survival of genes, and modelled the title of his book, *The Wisdom of the Genes,* on that of Walter Canon to transmit a similar message. This would mean that our evolution is conditioned by the equilibrium of genes and the organism with the environment. According to this hypothesis, the stability of the human body would result from dynamic laws, including the permanent mechanisms of genetic adaptation, being able to modify our capacities, without our will intervening. It appears to be a real functional adaptive determinism whose phenotypic expression varies according to the expression of genes. During a period of famine, for example, it is interesting to note that certain genes, hitherto unexpressed, will be expressed to facilitate survival. Upon resumption of a

---

8. Christopher Wills, *La sagesse des genes, Nouvelles perspectives de l'évolution (The Wisdom of the Genes: New Pathways in Evolution)*, Flammarion Editions, 1991. He explains how genes organised themselves in the past so as to adapt to changes. Christopher Wills opens the molecular "toolbox" of evolution to reveal unexpected gene specificities. Beyond our own comprehension of life, the genes seem to acquire an experience of life and a kind of "wisdom" during the billions of years which precede us.

normal diet the expression of these genes should be repressed. However, in some cases, the latter, for unknown reasons, could continue to be expressed in the offspring. This event would provide a logical explanation for overweight people inexplicably appearing in certain family lines, generating obesity in some children, even though they eat a balanced diet. In addition, it should be noted, if this hypothesis proves correct, that each new attempt to diet will not facilitate weight loss, but cause a defensive reaction of the body to reduce the weight loss by expressing the genes which support the transformation of proteins into glucose (gluconeogenesis).[9] This situation can cause despair for a patient who is making a major effort to lose weight by dieting, as these genes are expressed so much better than the reduction appears to be. This observation leads us to ask questions about the reasons which generate the expression of certain genes and those for which this expression continues to manifest despite the resumption of a normal diet. Does the body express an adaptive control reflex in anticipation of a possible modification of homeostasis, which would be caused by a famine whereas the latter would have been lived and memorised in epigenetic form by parents, grandparents or ancestors? Is this about a protective mechanism of genetic inheritance?

All the regulatory mechanisms of the genes, the immune system, the metabolism or the nervous system participate in our stability. Today, there are not many research teams in biology working on the subject.

---

9. Researchers Gilles Mithieux, INSERM (National Institute of Health and Medical Research, France) and Christophe Magnan, CNRS (National Centre for Scientific Research, France) respectively, have defined, in the small intestine, the expression of gluconeogenesis genes at the level of messenger RNA, of protein and of enzyme activity *Cahiers de nutrition et de diététique* (Journal of Nutrition and Dietetics), Vol. 41, NE4, September 2006, pp. 211–215.

In fact, two concepts of physiological adaptation can clash: one, considering that man has almost limitless capacities of adaptation; and the other, showing that man is conditioned by his terrestrial origins, the environment in which he developed coping skills, in particular genetic, during changes in the environment. In fact, the time scale gives us a mixed response: in the short term, our physiological adaptation of the terrestrial type, that is to say very limited, but in the long term it is not impossible to think we can survive under very different environmental conditions. Today, the notion of "physiological risk" joins with that of the "genetic risk", because in the short term we are able to modify our environment (e.g., atomic war). It is likely that our adaptation capabilities are so limited that, in the event of a sudden change in the environment, we could not survive. We previously noted that our stability, on which our identity depends, obeys laws of complex regulation.

Does this mean that we all look alike? If, in terms of species, there is no doubt; in terms of personality, it is not the same.

Each individual has a personality of his own. We speak of gifts, of capacities for a discipline: musician, poet, mathematician, physicist, doctor or soldier. This important concept of gifts is all the more significant as it combines the concept of the need for expression with the satisfaction of the various functional components of the body (cf. Chapter 2).

Even if the personalities vary and adapt to the trades chosen, certain individuals present behavioural predispositions. It is thus possible to anticipate certain individual reactions and certain character traits according to the expression of personal physiological characteristics. One century ago, medicine described a behavioural human typology in keeping with the operation of certain organs such as, for example, bile with sallow slender features and melancholic character, the kidneys with anxious and sad etc. We note that the stability or alteration of a state would generate identical expressions as much on a clinical level as on a behavioural level, translating a form of determinism of laws variously interpretable according to the expression of dominant or altered functions. The Cartesian relationship between cause and effect provides a logical understanding of the functioning of the body as also a benchmark which is reassuring for thought. However, the logical sequence is more than it seems, because the laws of regulation develop in a deterministic environment without one being able to really understand the final objective, if it is not the achievement of a state of balance. What is interesting is why, and the expression of this state of balance. Is this a state favouring reflection, life in society, making it possible to anticipate events, to feel free, responsible, to become self-aware?

If it is considered that our life is the result of a state of overall balance, it is understandable that Gregory Bateson[10] studied, at the school of Palo Alto, certain social mechanisms on a systemic level. Paul Watzlawick[11] studied the mechanisms of psychic communication between individuals. He referred to systemic analysis in order to analyse certain conditions such as schizophrenia.

---

10. Gregory Bateson, anthropologist, tried, across different disciplines to construct a cybernetic and systemic description of life. He is the author of a surprising ensemble, *Une unité sacrée, Quelques pas vers une écologie de l'esprit* (A Sacred Unity, a Few More Steps Towards an Ecology of the Mind), Editions du Seuil January 1996, for the French translation.

11. Paul Watzlawick, in relation with the work of Gregory Bateson, has applied the logical and cybernetic models to pathological (and normal) humans. Author of *Une logique de la communication* (A Logic of Communication) and *La réalité de la réalité* (The Invented Reality), he is one of the founders of the school of Palo Alto.

What seems obvious is that without this state of stability, we would be unable to grasp the metaphysical dimension of problems which we could not otherwise resolve. In other words, thought owes everything to stability, everything to balance, even if it comes up against the incomprehension of phenomena or with the absence of Cartesian explanation. Stability offers the capacity to think in a visible peace – a revealing state of grace of the happiness to exist in an unstable world. It lives and is felt with the passing days. Its respect, namely, that of the spirit of the laws, of "non-action" as recommended by Lao-Tseu in China, gives life a sense.

"The wise person is comparable with water, but with perfectly calm water."

"It is not in running water that men are reflected, but in still water; only it, by its stillness, is capable of stopping all who stop there (to be reflected in it)."[12]

This correlation between the notion of stability in a given environment and the ideal of life leads us to believe that we must respect the laws of nature while using them.

The stability of a state is seen relative to time, that is to say over time. We must understand that if we are not aware of the regulatory mechanisms that maintain our physical and mental status, we are witnesses to the ageing of the organism and its morphological changes. The stability of a state is not definite, but sufficient to understand the paradox of it. It is necessary to combine the impermanency of the states, since they all change, and the permanence of the laws of rejuvenation, of rebuilding. The determinism of the stabilisation laws does not mean identical results, however, we retain the same personality throughout life. This stability

12. Jean Grenier, *L'esprit du Tao* (The Spirit of Tao), *Tchoang-Tzeu 5*, Champs Flammarion, 1973.

does not symbolise a static state but suggests that we live these modifications which make us progress or regress. This observation has been integrated as a principal of thought in Buddhist philosophy.

The work of physiologists has remained within the framework of biology for a long time. Although the concept of physiological homeostasis has not been addressed on the basis of energy so far, physics (and more particularly electrophysiology) has become increasingly dominant in research on the human body. Besides multiple physical sensors serving to record physiological abnormalities, being able to determine disease – cardiac (electrocardiogram), cerebral (electroencephalogram) and muscular (electromyogram) – scientific research is increasingly interested in indications of energy produced in the tissues. Thus, Gamma-electroencephalography highlights areas in the brain where the appearance of energy varies according to specific stimuli (laboratory work of physiology of perception of Alain Berthoz[13] at the Collège de France and Jean-Pierre Changeux[14] on the brain at the Pasteur Institute). In fact, we have entered into the era of fundamental biophysics, which makes it possible to note on a physical level that which we currently identify biologically. However, these experiments do not play a therapeutic role and we can only note the effects of certain stimuli and try on an experimental level to physically modify their expression.

---

13. Professor Alain Berthoz has the title of Chair of Physiology of Perception and Action at the Collège de France (LPPA-CNRS UMR 7152). In addition he is president of the Institute of Biology of the Collège de France.

14. Jean-Pierre Changeux, neurobiologist, professor at the Collège de France is a member of the Academy of Science. He was president of Comité Consultatif National d'Ethique (CCNE) (The National Advisory Committee of Ethics for Health and Life Sciences) from 1992 to 1998.

If this evolution of scientific thinking is logical in itself, it is nonetheless disconcerting. Indeed, the biological equilibrium highlighted by Claude Bernard, and by the entire school of physiologists who followed, now tends to be studied from the point of view of the appearance of energy on a materialised given tissue. In other words, we are in the process of discovering the balance existing between the expression of energy within tissues and their materiality within the body. This approach necessarily arouses in us questions about the origin of the thousand-year-old Chinese concept of yin-yang balance[15] as our vision for stability evolves towards a vision of balance between energy and matter. Doesn't modern atomic physics define the energy-matter balance according to the equation of Albert Einstein: $E = MC^2$?

In the West, as in the East, it is impossible to speak of stability without incorporating the idea of regulation. However, can we identify these two theoretic approaches when they have emerged several thousand years apart?

Physical and Physiological Identity, Nutrition and Obesity

The physical identity of an individual presents a static, instantaneous, non-evolutionary character. It is concretely described as an identity photo which varies according to the stages of life. Who has not examined with nostalgia, old

---

15. In his book *Figures de l'immanence, Pour une lecture philosophique du Yi King, le classique du changement* (Figures of Immanence, For a Philosophical Reading of Yijing, the Classic of Change) François Julien explains the role of the traits of *yin* and *yang* and their distribution on change. It locates the origin of Chinese writing in the distribution of the three traits yin and yang within hexagrammes (superposition of six features).

identity photos where one's own image is gradually transformed and has taken shape imperceptibly? However, the morphological changes of an individual from birth to death provide dynamic evidence of the conservation of a stable, but not static, equilibrium. Despite all of these morphological changes, the memorised consciousness of the ego[16] persists in permanence. This individual recognition exists because of an adaptive memory, progressive and true to a principle of identity, recording the physical transformations and psychic evolution of the individual throughout their life. This would then mean that we integrate a vision of our body at once adaptive and impermanent, so much so that we experience difficulty in controlling our contours. Here, abstraction joins the concrete.

As for the physiological identity, it combines a dynamic, progressive vision of the body expressed through movement, and of attitudes related to a sequence of stable and harmonious images defining behaviour. It is necessary to differentiate the live photo of a tennis champion, and the film of a tennis match where the athlete's behaviour is analysed not only technically and strategically, but also in terms of his personality. The main point is that the physiological identity characterises the capacities of action. Yet each individual has specific functional predispositions which give him/her a stable originality, and will be able to guide the choice of a profession in society. Thus, to know an individual, it is necessary to know how he/she reacts.

---

16. Antonio R. Damasio, professor of neurology, neurosciences and psychology (University of Southern California) highlighted the activations of the cortical pathways and sub-cortical in recognition of faces and objects. He published *Le sentiment meme de soi: corps, emotions, conscience* (The Feeling of What Happens: Body and Emotion in the Making of Consciousness), Odile Jacob, Paris, 1999, p. 380.

My nutrition consultations have revealed the difficulty a medical nutritionist can experience when he wants to convey a message that is likely to change the behaviour of a patient. To be convincing it must be both simple and logical, explaining the 'why' of the behaviour. While treating the problem of overweight in particular, I have always refused to confine myself to a strict eating approach. To speak only of food and calories is not sufficient to resolve such a complex problem. I need to combine a set of components that affect weight gain.

Obviously, the first contact with the patient is essential. These subjects, often traumatised by disparaging remarks and their own self-image, have already tried many techniques to lose weight, hoping to find the doctor who will miraculously make them get thinner, if possible, without too much effort.

The consultation always proceeds according to the same principles: after having quickly explained to me the reasons for their consultation and the history of their weight gain, I note, as usual, employment status, family history of overweight as well as personal medical conditions and surgical procedures. A survey of the food consumed during the day provides the daily routine. Weighing, which I try to lighten up with an encouraging word, is followed by a brief referral to the body index – a statistical figure which indicates the level of overweight. I supplement this approach by taking measurements – an excellent way to compare between two consultations and to which the patients are much attached.

It is only after having given a report that we enter into the heart of the consultation. The aim is to clarify the physiological data and to accompany the patient during the day. Above all it is a question of explaining the functioning of the body, the role of food, what becomes of it in the body and the role of the environment. The word 'diet' has long been eliminated from my vocabulary. This term seemed to me to be at once restricting and insignificant. It also hampers more than helps the patient.

The objective is to make the patient aware of the permanent link the entire body has with the specific functional components. He/she must always have in mind a comprehensive view of the body. This approach is multidisciplinary. It avoids awakening a sense of guilt in a subject who tends to establish a relationship of cause and effect between food intake and weight gain. As I explained previously, the body is a self-regulating homeostatic system whose stabilising effects are also expressed in terms of weight. However, the body associates four interactive functional components, viz., metabolic, neuropsychic, immunological and eliminatory. Thus, food and mobility, analysis, central control of decisions and nerve transmission, immune defence system and elimination of waste represent the main functions of the body. This quadripartition of the circle makes it possible to make the role of food relative, while preserving its principal character. It should be considered that all functional components of the body are involved in a weight gain.

To provide an overall vision of the body by associating all these parameters during the same consultation is not simple. Moreover, it is necessary to integrate into all of this the influence of the environment on organic stability.

To objectively clarify all these data, I had the idea of drawing a circle divided into four equal parts, each intended to represent one of the functions which I mentioned earlier, i.e., the whole existing in the environment. I think, to modify the behaviour of an individual, it is advisable to separate and integrate a set of comprehensible scientific data in a circle. The circle, projecting a sphere which seamlessly blends all functional components, is a metaphor of the body adapted for the duration of a consultation and which facilitates memorising the advice to be followed. Thus, the circle can at once influence the patient and help him/her to progress. In addition, a benefit of this image is to help forget the complex caused by a guilt-producing view of the body. Therefore, I find it easier to compare the human body to a sphere whose dimensions vary according to weight gain. The representation of an individual, who grows bigger compared to a sphere of reference, is simply a circle with a larger diameter drawn on a sheet of paper.

The objective is expressed in these terms: to lose weight, it is necessary to reduce the mass of this sphere while acting on all of these functions. Returning to the page where I drew the circle and its four components, I draw a straight line in the lower part, and speak again:

> It is a mirror in which the image of the circle is reflected. So this is your physical representation in the mirror. What should surprise you is that you can look at yourself each day in the mirror without noticing any significant changes. This relative stability is all the more surprising given that there are important factors of food variations daily. On a physiological level, this state is called homeostasis, on a physical level, we talk about identity. In fact, you are so used to recognising you that it doesn't surprise you to see you unless you make some disparaging comments and that you notice yourself that there are little morphological changes. A pair of trousers which have become constricting, a coat which is too tight, a bulging shirt ... there is really something changing! However, even if your morphology is changing, it is unusual for this to happen within a few days.

As the demonstration is complete, it is essential to connect the components by making them interactive and to emphasise the role of internal and external stimuli on regulation.

Obesity does not generally appear suddenly. It is often progressive, sustained over time in a relatively continuous way, sometimes in increments. Some obese people are thus used to their image and may experience difficulty in changing it. They forge a personality around it and maintain it in a social context which they often enjoy. We sometimes have the feeling that they consider this situation like fate, independent of their will and behaviour, and for which they do not feel guilty. They know more implicitly that their state will not vary from one day to another. They learn how to live with their obesity.

This slow modification of the body is the result of the inertia of the body. I often compare it to that of a tanker which stops its engines long before the arrival in the port, driven by the inertial thrust. It is impossible to lose weight permanently without modifying the inertia of the body. Despite all the physical and psychological changes that we are submitted to over the years, our inertial system preserves our physical appearance, that is to say our identity. It is interesting to note that even if the quantity of food which we eat varies considerably from one day to another, our body does not necessarily reflect a significant gain or reduction in weight. With the principle of identity involving the recognition of morphological stability, we can say that all regulatory mechanisms constitute an inertial system which promotes weight stability.

To talk about stability to an obese person can appear insolent, provocative and disappointing. Indeed he/she sees the stability of his/her obesity daily. However, it is becoming clear that some obese people have a real predisposition for weight gain. To find logical explanations for their condition is still a challenge. If we conduct a family survey, we frequently find an obese father, mother, grandparents or other relatives. Heredity is a possibility which would be useful, but it is very difficult to determine the origin.

Should we conclude that it is impossible to change our constitution? This would signify, finally, that medicine is incapable of resolving the problem.

It is from observing the weight-gain of some subjects who had had a hyperprotidic[17] diet that I understood the link between a hyperprotidic diet and a starvation diet. The theoretic principle of a hyperprotidic diet consists firstly of facilitating the burning of body fat by cutting the intake of carbohydrates and fats while keeping an intake of protein. When the body has exhausted its reserves of glycogen from the transformation of simple sugars into glycogen, it draws its energy from fat reserves which are abundant since the subject is obese, eventually transforming protein into glucose. It will initiate a new reaction called gluconeogenesis which converts a part of the protein mass of the body into glucose. However, in a balanced diet, the body does not use this energy mechanism and this reaction occurs only in emergencies, in particular during periods of famine. It is completely conceivable that a starving body in an emergency, expresses some genes not previously expressed in order to transform constitutional proteins into glucose energy to sustain life if possible, or in other words, to help survival.

In developing countries, particularly in Asia, Latin America, the Middle East and Africa, some famines generate

---

17. Consumption of a diet rich in protein with the goal of achieving weight loss.

malnutrition whereas some famines generate a protein-energy malnutrition (previously called "protein calorie malnutrition" and "protein-deficiency") in the form of "marasmus"[18] or "kwashiorkor"[19] – extreme situations where there are significant electrolyte disturbances, source of oedema, and with a self-destruction of the body which consumes its own protein reserves, thus the muscles, and all organs composed of proteins. To start these reactions of gluconeogenesis, it is possible that certain dormant genes reactivate. However, with the resumption of a balanced diet, it is probable that in some cases (not in all cases), the expressed genes may not be repressed. This situation constitutes a glycogenic overfeeding of the body due to the addition of glycogen intake from normal food (carbohydrates) to which endogenous and exogenous proteins (gluconeogenesis persistent) whose effects on the pancreas will cause secondary hyperinsulinism, a necessary condition and sufficient for making fat and weight gain, must be added. This is why, it seems essential to me to guard a distance with respect to the hyperprotidic diet. Its genetic and toxic consequences, due to the nitrogen overload caused by proteins, are far from negligible on the body. It is a "famine diet" which profoundly modifies the expression of the genes and induces unusual physiological responses in the subject's equilibrium.

---

18. The concept according to which marasmus results from a calorie deficiency and kwashiorkor from a protein deficiency is simplistic because the causes are much more complex. There can be an intermediate state, the "marasmic kwashiorkor".

19. Kwashiorkor, discovered in 1930 by Cicely Williams of Ghana, is thus called according to a local name meaning "illness of the displaced child". Several metabolic changes occur and contribute to the onset of oedema. It is a deficiency of potassium and sodium retention associated with insufficient protein intake. Poisoning by aflatoxin, the presence of free radicals is an untested hypothesis.

Our morphological stability is conditioned by our eating habits and, in particular, by the quality of the food we consume. Any imbalance in the diet can cause biological imbalance resulting in anomalies of the biological constants. It is thus impossible to speak of stability without discussing the notion of the biological norm. The body mass index (BMI), calculated by gender, height, weight and age, gives an indication of the theoretical ideal weight. The formula of Lorentz,[20] frequently used, takes into account the height and gender of the person and not age or the different morphologies (long-limbed, slender, average, stocky, broad, etc.). Thus, Lorentz's formula gives an appreciation of the ideal weight at a given moment but does not consider the history of obesity and the appearance of pathological risk linked to weight gain.

Biological standards are relatively easy to quantify. The blood formulation expresses the biological status of an individual. For example, the glycaemia measuring the amount of sugar in the blood varies depending on the hour at which it is measured. Varying between 0.70 and 0.90 between meals, its normal value one-and-a-half hours after the start of a meal (postprandial glycaemia) should not exceed 1.50g/l. The glycosylated haemoglobin, or glycated haemoglobin (HbA1c), is a retrospective and cumulative index over two months of the level of glucose in the blood. Its normal value is usually understood between 3.5 per cent and 6 per cent. If the glycated haemoglobin is not sufficient to establish a diagnosis of diabetes, it is nevertheless witness to the average of the glycaemia and adhesion of glucose on proteins.

---

20. Men: Ideal weight (in kg) = (Height – 100) – (Height – 150)/4). Women: Ideal weight (in kg) = (Height – 100) – (Height – 150)/2.5). For men, as for women, the height is in cm.

On a morphological level, it is more difficult to set standards. On the other hand, it is possible to establish a link between the occurrence of certain pathologies and obesity. Consequently, in a determined environment, obesity can represent a pathological risk which is statistically established.

The variations of weight over the course of the lifespan give many indications on physiological imbalance of the organism and constitute a pathological risk factor (diabetes). This information concurrently guides treatment on the one hand, and on the other hand, and especially, the behavioural control of the subject concerned.

The Sinusoidal Law of Evolution

Any morphological change of an organism calls for a modification of the genetic information, whether it is related to a mutation, punctuated or not, or with the expression of new genes, some of which may have been put in reserve. The non-expression of certain genes remains an enigma today promoting the emergence of various hypotheses.

Take the example of the horses abandoned to live in the wild in a desert at the end of the First World War, and which developed during the following years, in a few generations, a system of saving water which enabled them to adapt to life in an arid area and to survive. The fabrication of this loop of regulation, highlighted by anatomopathological examinations, required new genetic information which has emerged in the space of a very short period (some generations).

Are they dormant genes which would have appeared a long time ago in an identical situation (desert environment) or of a "punctuated" change caused by the sudden change of environment? In the first case, a "non-expressed" gene would come into action following the environmental changes, which does not explain how this gene would have appeared in the beginning. One possibility could be that, originally, these genes were developed gradually taking into account a slow environmental change. During a more abrupt climate change (transition from a hot climate to a cold climate), they would have stopped being expressed and would have been put in reserve. This could explain the rapid reactivation when the horses were abandoned in the desert.

Could we speak of a genetic regulation like that which exists for thermoregulation? The example of gene expression during times of famine encourages me to think that it is more of a re-expression of genes than a punctuated change.

Conversely, can a gene stop being expressed following a change of the environment?

With environmental conditions varying, some genes could lose their expression, become silent and be stored as a kind of "event-driven memory," ready to reappear when needed. The expression and non-expression of genes would translate a regulatory system of the body supporting the protection of the genome during environmental changes.

In reading *The Invented Reality* by Paul Watzlawick I realised the capacities of adaptation of certain animals such as the dolphin which, after having been a land mammal, returned to the water:

> If we admit that the dolphin returned to the sea millions of years after it had adapted to life on land … the question of knowing why this creature is equipped with such a large brain still remains unanswered …. They returned to sea and thus abandoned a certain evolution, which we must consider, until evidence to the contrary is

> presented, as fundamental to the appearance of a civilised race. The hands of dolphins have mutated into flippers (their skeletal structure is still that of a hand) ...[21]

A phenotypical regression would have proceeded on the level of the limbs, transforming them into fins. Why did these terrestrial animals return to the water?

Is the adaptation of the organism to different environments synonymous with progress, does it translate as an improvement of the total capacities or is it a physical adaptation to the ambient environment? It would be the function of balance between the organism and the ambient environment which would create the body! Is genetics enough to explain these phenomena?

For many years, it was suggested that the sequence of a molecule of DNA represents the only basis of heredity. However, in recent years, it appears that there is another form of storage of heredity based on epigenetic phenomena.

Since genes alone cannot explain the complexity of life, some researchers are interested in the process providing the explanation of the links between genes and the environment, namely, the epigenetic phenomena. The term 'epigenetic' defines the transmissible and reversible modifications of gene expression not accompanied by changes of the nucleotidic sequences. This type of regulation can target the DNA, RNA or proteins and act on the level of the core or the cytoplasm. Epigenetic modifications constitute one of the foundations of biological diversity. This trail resulted in thinking that besides our genetic inheritance, we also

---

21. Paul Watzlawick, *La réalité de la réalité, (comment le réel est réel), confusion, désinformation, communication (The Invented Reality* [How real is real?], Confusion, Misinformation, Communication), Editions of Seuil, 1978, p.157.

transmit to our descendants characteristics acquired during life affected by the influence of the environment.

The theory of epigenesis creates a field of freedom of genetic expression in opposition to the deterministic hypothesis of transmitted genetic inheritance. Many people agree that there would be no learning without adaptation mechanisms. Jean-Pierre Changeux defined epigenesist as a "combinational mechanism no longer involved in modifying the genetic material, being exerted at the level of all nerve cells. It has its origin in the topology (that is to say, the geographical disposition) of the network of connections which are established between neurons during development". Changeux highlights this mechanism when he describes the plasticity of brain nerve cells useful for learning and the adaptive memory in relation to the acquisition of new communication channels. We cannot attribute the sole responsibility of brain development to the genome. The brains of twins coming from a single egg fertilised by one sperm (identical twins) turn out to be different. The cerebral convolutions are not distributed identically and their phenotypical variability would prove the influence of the environment on their development.

In humans, the mechanisms of adaptation to the environment, by eliminating the expression of specific genes, could lead to a phenotypical transformation and temporarily restrict certain human capacities. Thus, our genome could generate, during profound changes in the environment, morphological changes while maintaining the same genetic potential, part of which would be set aside without us having any decision-making power. A visible and functional morphological regression could occur, followed by a slow return to the previous balance in accordance with the laws of evolution.

Indeed, does an ideal environment exist which supports the development of human capacities, and another which is harmful where man would lose some of his capabilities and current identity?

Will all the changes we currently subject the planet to affect our genomes to the point of generating a non-expression of some genes? We would not talk about evolution in this case, but of temporary regression!

3

Physiology and Regulation

Social Repercussions

Modelling the Functional Components of the Human Body

Before penetrating the world of pathology, medical studies teach fundamental sciences. Among those, physiology occupies quite a particular place. In the *Dictionary of Medical Terms*, the authors Garnier/Delamare[1] define physiology as "part of biology, which aims to study the properties and functions of organs and tissues of living beings".

It can be classified into cellular physiology and general physiology. While cellular physiology studies vital phenomena localised in the cell, general physiology studies the functioning of the body.

1. *Dictionnaire des termes de médecine* (Dictionary of Medical Terms), 22nd Edition, Garnier/Delamare, Maloine, 1989.

Diagram 1

Homeostasis and Identity

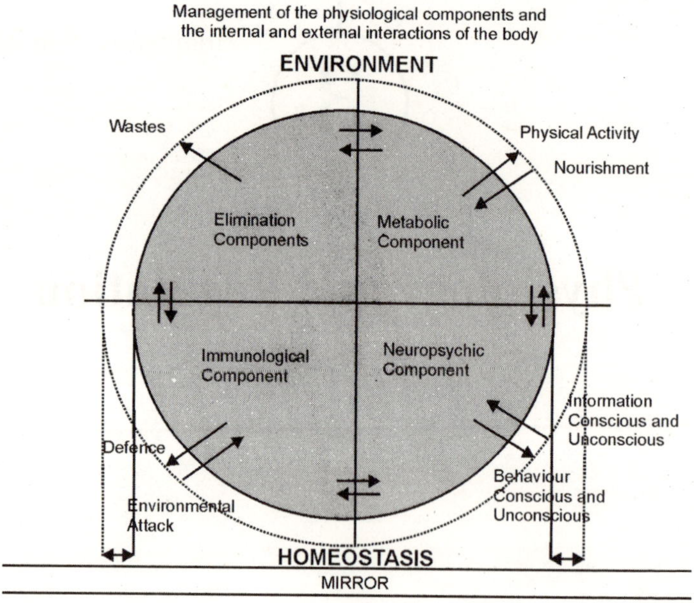

Due to the scope of its field of intervention, physiology links the most intimate mechanisms of biological regulation with their behavioural expression. For this reason it is logical to consider that it bridges the gap between experimental science and philosophy of action.

To think, choose, eat, drink, sleep, defend oneself and eliminate express within the same body the inseparable and essential functions to maintain a state of balance (homeostasis) and ultimately, survival.

The term 'homeostasis' was employed for the first time in 1929 by Walter Cannon[2] in the book titled *The Wisdom of the Body.*

---

2. Walter Cannon, *La sagesse du corps* (The Wisdom of the Body), Paris, Editions de la Nouvelle Revue Critique, 1946.

Diagram 2

Extra-organic Functional Prostheses

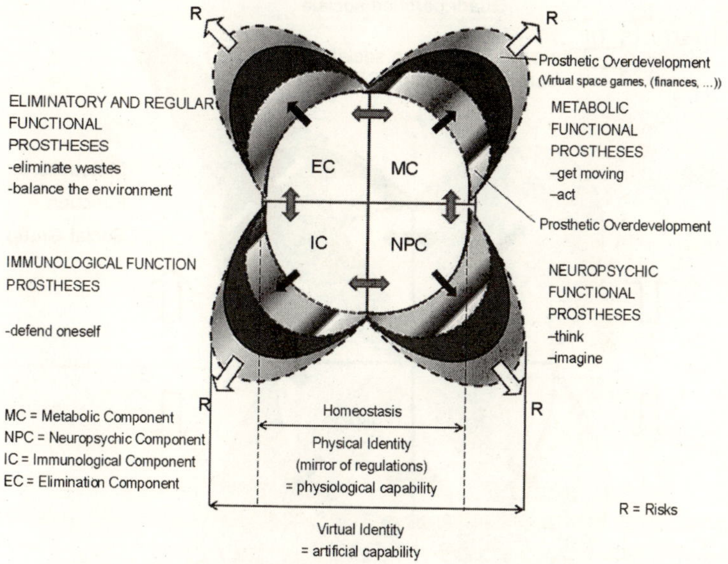

It was in reading the *Handbook of Human Physiology,* written by Jack Baillet and Erik Nortier, prefaced by Professor Roger Guillemin, Nobel Prize winner of Physiology and Medicine, that I realised all the homogeneity given by a homeostasic approach of physiology "in the land of Claude Bernard".

The work of Norbert Wiener[3] on the modelling of the control of an action directed according to the flow of information has made it possible to apply his theories to physiology. As it is expressed by Jack Baillet and Erik Nortier,

3. Norbert Wiener (1894–1964) was an American mathematician, theorist and researcher in applied mathematics. He was the founder of cybernetics, a science which formalises the idea of feedback (retroaction) and has implications in the domains of engineering, control systems, information technology, biology, psychology, philosophy and the organisation of society.

Diagram 3

Composition of Functional Social Groups

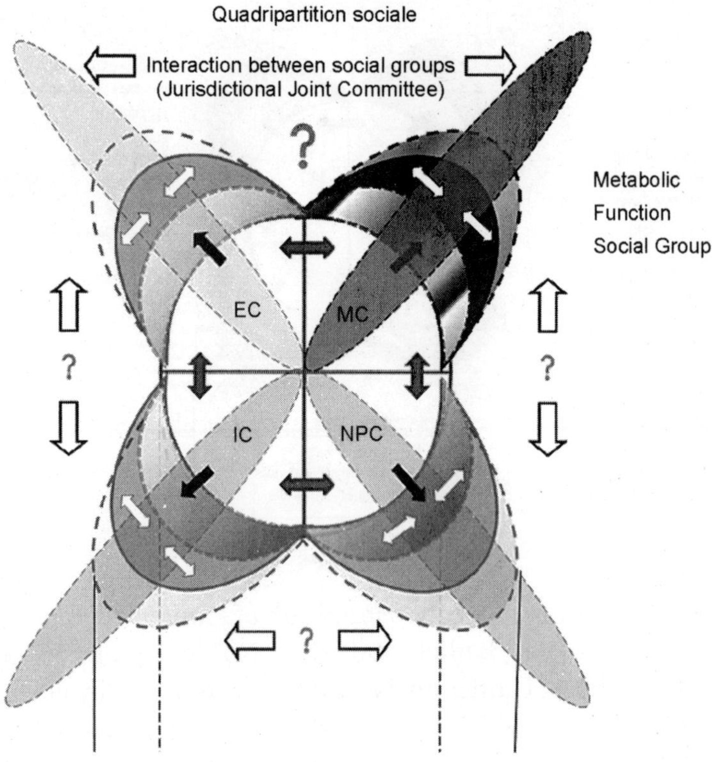

"Cybernetic modelling will thus consist of applying these processes to physiology of the systems which 'control' homeostasis."[4]

Gregory Bateson, Jack Baillet and Erik Nortier then supported the idea of modelling of Norbert Wiener by extending the study of behaviour: "The major interest of a cybernetic system is to result in a usable topological model

4. Jack Baillet and Erik Nortier, *Précis de physiologie humaine* (Handbook of Human Physiology), Ellipses, 1992, Editions Marketing, p. 30.

which makes it possible to suitably classify the physiological or physiopathological facts."[5]

Our topic concentrates on the development of the four major physiological functions of the body by tools which I have named "extra-organic prostheses" (EOPs) or overdevelopment prostheses. If they express the functional and complimentary diversity of our physiological capacities, they are, however, interdependent and interactive, self-regulating with respect to a state of balance, defining a state of dynamic stability (the steady state described by Walter Cannon), which facilitates modelling of the whole.

Each physiological component generating needs accomplishes specific functions. Several major, essential needs must be met to maintain the stability of the whole. The body functioning as an independent system in the environment draws its energy resources (food) there, integrates information, defends its identity to survive and eliminates the waste of the products which it used and catabolised.

We will now succinctly study the role of each functional component before analysing their artificial extension and the risks incurred.

Metabolism describes the ensemble of biological operations necessary for cellular life. It defines the ensemble of transformations of foods used for the maintenance of cellular life, it is as equally about the contribution of organic substances (anabolic) as the elimination of waste products (catabolism). The main objective of metabolic function is to nourish the body in order to make it dynamically autonomous to gain mobility and the expression of our capacity for action. It is related to food production, which is itself very dependent on economic infrastructure.

---

5. Ibid., p. 31.

The book of Baillet and Nortier (already cited) makes it possible, with its continuation, to establish a connection between physiology and behaviour, by considering the social impact of our physiological functions, particularly on the neuropsychic level. From the anatomical description of the brain, it is necessary to study the behaviour from an evolutionary perspective. Thus, in the 1950s, Paul Mac Lean proposed for man the image of a triune brain with a reptilian core, a limbic brain (specific to that of the old mammalian) and finally a cortical system of which the "irrational" frontal development made the 'Homo' a radically changed primate.

The *neuropsychic* function controls conscious and unconscious behaviours.[6]

"It is obviously a challenge to imply a strict correlation between the neurophysiologic systems, supported by centres, often intermediary or effectors of multiple behaviours and our conduct."

The lower axial system corresponds to the brainstem of anatomists with, above the spinal cord, the stacking of the bulb. The emotional afferents are manufactured by the axial system that modulates cortical functioning. It allows survival by modulating breathing, heart activity (by stabilising the blood pressure) and by regulating the digestive functions. The hypothalamus is part of the higher axial system. It manages certain homeostatic functions and is also involved in both endocrine control and the survival of the species. Instinct, learning and memory can be modelled on the basis of the animal model to lead to what they call "human software".

"To address studies of human behaviour within the framework of a precise physiology, even if everyone accepts

---

6. Behaviour: an act conducted directly or indirectly in order to satisfy a need.

an animal genealogy and the stability of human behaviour for thousands of years, is to expose the unwary to crossfire: on the part of scholars specialising in commentaries on human beings, and the 'real' physiologists who consider these problems as being outside a possible scientific approach."[7]

The *immune* function of the body is identified as a homogeneous and autonomous defence system guaranteeing the protection of our physiological identity against internal and external attacks (bacteria, virus, etc.) to preserve our survival. To realise that each individual possesses an immune system is to immediately think of the fight to stay in good health. The systematic vaccination of children and adults, and equally the researchers of vaccines to fight against diseases such as AIDS, the H5N1 virus and various cancers, contribute to this fight which constitutes a preoccupation of our society. In diffusing information about the progress of research in this field, public authorities make everyone aware of the role of the immune system on the balance of the body and, consequently, on good health. In addition, management of the immune system through the fight against free radicals, poor dietary habits, lack of physical activity and lifestyle has become common practice. For that matter, medicines called "anti-oxidants" are sold over the counter in pharmacies!

In facilitating the elimination of wastes which are toxic substances and likely to destabilise the harmonious functioning of the entire body, some organs – intestines, kidneys, lungs, skin – fully support homeostasis.

---

7. Jack Baillet and Erik Nortier, op. cit.

The blood, respiratory and nervous systems promote the synergy and specificity of exchanges between the functional components and the environment.

In the end, all these components function simultaneously and are interactive. To highlight the interaction of the different physiological components of the body, it is interesting to study, in athletes, the problems[8] posed by thermoregulation as well as their expression. The significant production of heat linked to muscle metabolism (metabolic component) in long duration sport events results in an increase in core temperature (neurophysic component). This hyperthermia is accompanied by dehydration, which causes a sharp decompensation, tritely called "heat stroke". Sweating (elimination component) represents the main component of thermolysis. Subject to central regulation, the rate of sweat secretion which can be as much as several litres per hour is proportional to the quantity of heat which must be eliminated to assure homeothermy. The evolution toward a serious hyperthermia is due to impairments in the functioning of the hypothalamic thermoregulatory centre (neuroendocrine system) and the drying up of sweat secretion. Having drinks spontaneously during the event motivated by self-assessment of needs cannot fully compensate for the water loss. Specific instructions about the timing of water intake promote improved performances.

By this simple description, we see the complementary nature of the functions and the multiple interactions of physiological components of the body respecting its stability in a dynamic action.

---

8. P. Vanuxem, E. Fonaris, N. Makhtar, N. Amrani and F. Commandre, *Les probleme de thermorégulation chez les sportifs* (The Problems of Thermoregulation in Athletes), Fac. Médecine lab. Physiologie respiratoire, Marseille 13385, France, Médecine d'Afrique Noire, 1990, Vol. 37, NE 11, pp. 686–688.

Extra-organic Prostheses and Risk Visibility

A prosthesis is an artificial device intended to replace a limb, an organ or a joint. Widely used in orthopaedic practice, prostheses have diversified and today concern all parts of the body. By extension, today, we can speak of prosthesis to designate a tool replacing man's action, that is to say, supplementing his functions. The awareness of his own limitations has led man to seek ways to increase his capabilities of action and at the same time his autonomy. He has developed tools adapted to the four major functions previously described. While being completely independent of the body, these functional prostheses will also artificially increase his autonomy and his power of action because of the functional objectives which they represent.

Man has designed increasingly powerful production equipment. Firstly, he extended his hand and maximised the metabolic function (his muscular function) thanks to the spade, the pickaxe, the plough, etc. Since the invention of the wheel, he has reduced his workload, using the wheelbarrow, the bicycle, etc. Then, the invention of the steam engine, or internal combustion engine, allowed him to replace muscular function with machines. Today, the prostheses complimenting metabolic function are described under the name of metabolic extra-organic prostheses (EOPs) or overdevelopment prostheses. At the same time, they concern the management of metabolic needs (production of food and its consumption) and the prosthetic use of metabolic function (construction of automobiles, trains, aircraft, etc.) which increases our capacity for action, particularly our mobility and scope of action.

The production of new tools requires an adapted infrastructure to manage the desired production from A to Z: a hierarchical social organisation (company employees, managers, project superintendents, workmen) centred on the production of EOPs will be interacting with 'consumers' likely to buy the 'products'. The whole thing is organised on a systemic level to meet very specific 'needs' and obey the law of supply and demand.

A socially diversified and differentiated group will also provide assistance to enable the company to achieve its objectives, whether they are designers, developers, economists or financiers, who propose, fund, analyse and compare the production of the EOP. Thus, a complex social organisation is established around each EOP, with the purpose of achieving its realisation.

In this perspective, we see a group of individuals with very different but complementary trades gathering together and whose objectives converge – to build metabolic prostheses.

It is impossible to analyse all the risks generated by the establishment of these EOPs in a few lines. However, risk analysis can be done from the conception of the metabolic EOP right up to its production by taking into account the various interactions – pre-production and post-production – the design of the EOP will be confronted with market research and the ethics of need and supply and demand. In the beginning, the exploitation of primary materials, such as wood, petrol, production of electricity by nuclear power centres, minerals, etc., can result in economic, social and environmental instability, which can be detrimental. Moreover, politically speaking, this thirst for exploitation can pose the risk of further weakening of some poor countries, which believe in miracles or the mirage of their own development. The choice of location for the manufacture of certain products will have to involve the assessment of its environmental impact. It is necessary to study the entire production line for possible risk implications, whether they are direct, indirect, environmental or social. Finally, down the line, the manufacture of the EOP within the company will generate waste materials which will have to be treated.

Representatives from the metabolic centre currently tend to impose their will on the markets, sometimes in contradiction with the national and international legislations in force, which establishes power relations whose effects are difficult to evaluate in terms of the risks incurred. Take the case of some petrol tankers or container ships which should not still be in service because of their dilapidated state. Some consortia, while engaged in getting maximum profitability on their transportation, do not care about the consequences of a shipwreck and the pollution it generates. Unfortunately, if an accident occurs, such as the sinking of the Prestige on the Spanish coast, we can not only assess the effects of pollution on environment and health, but also judge the difficulties in making the damage identified and obtaining compensation, which in any event, and whatever amount, will not solve the problem of the irreversibility of some of the damage. An international court still has a lot of difficulty in imposing its authority.

The financial crisis of 2008 highlights two parallel, interactive worlds, which are expected to mingle as well as diverge – that of finance and production. The lure of profit drove some finance companies to take such risks that they call to mind the behaviour of casino gamblers rather than that of financial managers responsible for their customers. In the absence of actual collateral, some financial players fabricated the financial guarantees of insolvent people in order to increase their earnings, making the entire banking system vulnerable in an economic crisis like that of the subprimes. Subsequently, there is the blackmail of players who lose and who do not want to be ruined. Without sufficient external control, the financial system could develop in a hypertrophic manner, and that to the detriment of all investors and, more broadly, the entire business world. Faced with the lure of profit, the responsible financial authorities ignored the risk taking with a thinly veiled hypocrisy, knowing it could ruin the investors and companies which trusted them. Faced with an obvious gap in the law, some unscrupulous financiers have seen fit to assure their own unique advantage, to be above the rules and to thus ridicule any ethical charter.

In conclusion, contemporary civilisation has woven a socially diversified social fabric in order to manage these metabolic prostheses – a delicate task because of the direct and indirect risks, and an absence of control systems required to perform this function. A court established by experts in economics and finance (belonging to the metabolic centre), knowing all the workings of these financial companies, is only able to slow the growth of these virtual companies. By distorting the world economy and ruining the poorest people, they continue to enjoy a privileged situation founded on blackmail. Officials of the financial world have entered a universe similar to the Mahabharata[9] which tells how the fate of the kingdom was decided by a game of dice: the eldest, Pandavas lost everything to the Kauravas. Has the time come when the fate of the world will be decided by a game of dice?

The neuropsychic EOPs serve to support thinkers, whether philosophers, religious scholars, researchers, poets,

9. *Le Mahabharata, la première partie de dés,* Livres II (The Mahabharata, The First Game of Dice, Book II), 57, 60, 62–65, Flammarion, p. 196.

journalists, politicians, etc. In terms of research, an institution like the Collège de France evokes the role that these thinkers hold in our society, all chairs together. For them the books are a database which I compare to a neuropsychic prosthesis, supporting the expression of research combining fields as diverse as mathematics, physics, chemistry, biology, religious history, philosophy and all literary subjects, creating currents of thought constituting humanism in its noblest representation. Ever since François 1st and Guillaume Budé, the College de France, like the prow of a ship, has furrowed through every canting seeking to perceive, explain and thus shed light on human existence. Today, neuropsychic EOPs are represented by all types of information, whether it be books, newspapers, radio, television, telephone, internet – in fact, all means of communication that facilitate the transmission of ideas.

The computer is unquestionably the most revolutionary neuropsychic prosthesis of our time. Originally designed as a very specialised instrument facilitating complex calculations, it has become, through its capacity to store data in a limited space, a part of our memory which, thanks to modern means of communication, is able to interconnect one with every country in the world. Thus, it has profoundly changed the cultural landscape of all nations. Facilitating information exchange through networking (Web), the computer has become the communication system of modern society to such an extent that sometimes there is concern about some people being addicted to it. By conveying a wealth of knowledge, the internet constitutes a global database for better or for worse. However, the image transmitted on the screen creates a virtual world, which combines genuine information and lies, mixing dreams and reality, sowing confusion in some minds. Its effects on human behaviour and mental health are far from being evaluated at an appropriate level (risk of schizoidism[10])!

The neuropsychic EOPs have already been used frequently in the 20th century to manipulate information. If the extraorganic neuropsychic prostheses constitute real progress in information exchanges, it is no less true that they could become a tool to interfere in people's lives, to spy on habits, to become an instrument of power in the hands of those who wish to influence the whole of the society in its choices. The neurophychic EOPs support the interplay of politicians who try to sell their image to seize or maintain power. Indeed, the impression transmitted in all its virtuality gives the individual a considerable potential power, which is enough to win over the public. Television brings us daily evidence of this. The power relations are so much stronger that some may use other prostheses to their advantage, especially the metabolic and immunological EOPs.

The neurosensory EOPs developed particularly at the end of the 20th century and today play a predominant role in

---

10. Schizoid personality disorder: A pervasive pattern of detachment from social relationships and a restriction of the variety of emotional expressions in relationships with others, which appears in early adulthood and is present in various contexts, as testified in at least four of the following demonstrations: the subject does not seek nor appreciate close relationships including the intrafamily relations: they almost always choose solitary activities, have only a little or no interest in sexual relationships with other people, experiences pleasure only in rare activities, otherwise not at all, has no close friends or confidants, apart from his/her parents of the first degree, seems indifferent to the praises and the criticism of others, shows coldness, detachment, or other blunting of the feelings.
Manuel diagnostic et statistique des troubles mentaux (Diagnostic and Statistical Manual of Mental Disorders, American Psychiatric Association), DSM-IV, Paris, Masson, 1996.

the population. Indeed, they were popularised because, by generating needs, they became the object of a trade whose expected profits are connected with their development. Let's say, in a metaphorical way, that they are located in the periphery of the neuropsychic component, and that they are interactive with the environment. This zone, dedicated to communication, represented by the organs of the senses, contains all the detectors of the environment. It is intended to inform the body about changes in the external environment to enable it to adapt consciously and unconsciously. Whether interactive hearing prostheses generated by various means of oral communication (telephone and mobile phone, radio, MP3 players, etc.) or visual prostheses (advertising displays of paper or light, film, television, etc.), everything is focused on creating new needs satisfied in a virtual world. James Cameron, a former physics graduate of the University of California, and director of the film *Titanic*, has been working for years on a new project intended to revolutionise 3D movie theatres. He also produced a science fiction movie, *Avatar*, which was released in December 2009. He points out: "When you watch a scene in 3D, things appear even more real. The visual cortex is tricked, your brain doesn't distinguish the difference, it thinks that what is sees is real. 3D results in a total immersion, you have the feeling of physically being in the film and taking part in its history."[11]

Shouldn't we worry when the acknowledged aim is to confuse the virtual with reality?

In 1990 Paul Verhoeven adapted for cinema under the title of *Total Recall* the short story of Philip K. Dick. We can

11. Jean-Luc Wachthausen, *"James Cameron, le visionnaire du cinéma"* ("James Cameron, The Visionary of Cinema"), Le Figaro, 26 December 2008, p. 21.

recall it for you fully. Set in the future (2048), it follows the life of a physically normal individual who dreams of going to the planet Mars but who is not rich enough to pay for the trip. In spite of the dissension of his wife and friends, he goes to a company which has invented a machine that can implant the memories of choice into the minds of its customers. Following a planned operation, the hero becomes a secret agent, eliminating his wife and all his old friends, goes to Mars and loses his personality by changing his name. Leaving the cinema hall immediately after seeing the film, some viewers believe that the hero actually lived the events which took place before their eyes. However, the manipulator explains before the surgery that everything has been designed so that the evolution of events is connected in the dream as naturally as possible. If we do not assume that the hero lives a dream from the moment he was taken over by the "Total Recall", we too lose the concept of truth and forgery. We no longer know what corresponds to pure delusion or to reality.

Since all these prostheses endeavour to make us confuse dream and reality, aren't they making us lose our bearings? This form of addiction to the virtual is similar to drug addiction. For pleasure, an individual loses control of his/her behaviour and lives alone in a daydream. Loneliness, isolation and the search for love – aren't they the engines driving these behaviours? Memory built on virtual images and sounds scrambles our points of reference: it presents to us Claude François*, Daniel Balavoine* and many others as if they were still living. We don't know any more! By charging the short-term memory with increasingly old references on a repetitive mode, the mind sterilises its inventive genius and creative power. It is necessary to know how to, and to be able to, isolate oneself in order to create!

*Very known French singers died many years ago.

Yet, the tools provided by these prostheses are fabulous as they facilitate communication (the example of the mobile phone is significant), allow access to information and culture in isolated regions, help fight against ignorance, can be used to support medical treatment, etc. For example, we cannot help but see the major benefits of flight simulators to learn to fly, but how can we fail to have, simultaneously, an ulterior motive for its possible uses?

It is necessary to establish ethical rules in relation to the individual and collective risks generated by sensory EOPs. This will lead to a new jurisdiction which must involve paediatricians, general practitioners, psychiatrists, neurologists, psychologists, sociologists and economic players. As of today we perceive the risks of psychological imbalance that our society sometimes unconsciously builds in the excitement caused by new discoveries. Let's not underestimate the risks!

The immunological EOPs were originally designed to help in the defence of the individuals and assist their survival, i.e., to physically protect them from attacks. Clothing would be, according to this approach, a first protective system of the body against the cold. To survive, man had to hunt: thus he armed himself. Very soon he sensed the need to defend his territory and used offensive weapons against his fellow men. Some organised themselves to defend their society and developed a social structure which grouped individuals who had the physical and psychological predispositions to achieve this responsibility. Each nation developed individually with a dominant culture seeking to impose itself within its territory – thanks to men who had the responsibility of ensuring the protection and defence of their country by organising themselves into a hierarchical structure – the army.

Today, the risks generated by the immunological EOPs are in keeping with the power of weapons of mass destruction. They represent a veritable "Sword of Damocles" hanging over the world. Concerning defensive weapons, almost all have become offensive, capable of creating terror overnight. Any force called "deterrence", although justified intellectually, represents a real danger in the sequence of conflict. Today, armaments have become a tool of blackmail, all the more disturbing now that they are in the hands of independent "small armed units" threatening entire nations. This is the problem of terrorism. If the army is representative of the organisation responsible for the defence of the nation, terrorists constitute a force all the more dangerous. The whole group of prostheses, metabolic, neuropsychic and immunological – which are placed at their disposal, increases their capacity for action considerably and facilitates their destructive power.

Finally, EOPs associated with waste disposal were belatedly dealt with. Since the middle of the 19th century, the industrial society has been discharging toxic waste. It was only at the end of the 20th century that ecologists worried about the implications of the existence of this waste in the environment. Today, an effective prevention policy is formulated because, in view of studies and climate change, the warning signs are flashing. It should help slow down and control the proliferation of waste discharges, such as radioactive, chemical or biological materials, whose toxicity can be dangerous both in the short and long term, causing cancers, allergic reactions, cardiovascular diseases, etc., and requires their care by specialists in the techniques of reprocessing and disposal. A social organisation was formed in order to manage all these problems.

It is in this context that ecology has become known. However, few people, apart from the experts, present it as a science. "It is tempting – and it is the position of many ecologists – to think that everything comes from a confusion between science and militant speeches."[12]

It is James Lovelock who developed the Gaia hypothesis following comparative observations between the atmosphere of Mars and that of the Earth. He was particularly surprised by the consistency of the gaseous composition of Earth's atmosphere despite climate change. He hypothesised that a regulatory mechanism favoured terrestrial homeostasis. He named this explanatory vision of terrestrial balance the 'Gaia hypothesis', named after the Earth goddess. As quoted by Jean-Marc Drouin, James Lovelock does not identify his approach to that of an ecologist:

> The difference between the Gaien concept and the ecological concept of our planet stems in part from their history. The origin of Gaia was the vision of Earth from space, which revealed the planet as a whole and not in detail. Ecology is rooted in basic natural history and in the detailed study of habitats and ecosystems without worrying about the big picture. One is not able to distinguish the trees from the forest; the other, the forest from the trees.[13]

Depending on the country, each of these prostheses may become overgrown or degenerate; and each case may lead to social imbalance, generating problems and eventually conflicts.

The financial crisis of 2008 is an indication of the instability existing between the reality of work and the inability

---

12. Jean-Marc Drouin, L'ecologie et son histoire (Ecology and Its History), Flammarion, 1993.
13. J.E. Lovelock, *La terre est un être vivant: l'hypothèse Gaïa* (The Ages of Gaia: A Biography of Our Living Earth), Monaco, Ed. Rocher, 1986 (ed. or. 1979), p. 148.

of financial institutions to judge the reality of a situation. This society produces speculators who, armed with information conveyed by computers, gamble on share prices, ignoring the value of the actual work of those who produce "that" with which they speculate. The crisis of the subprimes[14] could have been prevented if insurance companies had held their role of insurer.

Ulrich Beck, in his book titled *The Risk Society*, notes: "Class inequalities and social inequalities in relation to risk may overlap, be determined and generate each other".[15]

Today, conflicts cause fear, because everyone knows that a slip could lead to a global disaster.

While following the progress made in the formation of EOPs, we realise that each function develops a specialisation supported by a specific social organisation in relation to the objective researched. Each organisation leads to a form of independence and autonomy which can finish by excesses. Consequently, some questions arise: How to regulate each sector? Do today's politics have the capacity to counsel, to limit or forbid certain production? How will the risks be managed? Is it possible to escape the inertial phenomenon of functional specialisation? What are the relations between all the social organisations? Is there synergy between them? By which methods can it be fostered and developed?

All things considered, the evolution and representation of the various functional prostheses in our society must

14. The crisis of the subprimes (subprime mortgage crisis) is a crisis which affects the sector of American mortgage deeds to insolvent people practised by the Federal Reserve Bank (FED) between 2003 and 2004.

15. Ulrich Beck, *La société du risque, sur la voie d'une autre modernité* (The Risk Society: Towards a New Modernity), Champs Flammarion, Paris, 2001, p. 80.

encourage us to ask ourselves the question of their specific, individual and collective management. It leads ultimately to carry out a political debate on how to run a modern nation.

Thus, the management of the four major human functions centred on mobility, thought, defence and the management of elimination is inseparable from that of the prostheses which influence physiological balance. It is essential to maintain our autonomy, our freedom of action by creating a system of risk management whose purpose will be the facilitation of the balance of mankind.

In light of these considerations, it appears that man, by significantly increasing his capacity for action through EOPs, is now able to alter his homeostasis.

High-Risk Society and Low-Risk Society

Another objective of this chapter is to use physiology as an analytical tool to define high-risk and low-risk societies. In this context, we will study the influence of human behaviour on the physiological balance of the individual in his/her environment. We will also focus on the concept of physiological risk. Having already highlighted four major human functions essential to the survival of the reproductive system, it is interesting to analyse the causes of homeostatic instability, in other words, of this stable condition (steady state) which characterises us.

The creation of prostheses specific to each function gives the individual a power which he/she only partially controls. Thus, despite all the precautions, cars produce not only their share of fatal accidents, like planes, trains or boats, but also sea and air pollution, having secondary effects on man. These inventions spawn primary and secondary risks which are difficult to control. Yet, the effects of these risks on the environment and on human physiology are increasingly characterised, whether by the greenhouse effect, allergic diseases or cancers, etc.

Many people accuse man of being responsible for the development of environmental risks and agree that technological developments should be monitored and controlled by groups capable of measuring the effects on ecological and human balance. This awareness is expressed in daily newspapers and in a more specific fashion in the works of ethics committees such as the CCNE[16] created in medicine. Ethics relates to all forms of human activity and now goes beyond the field of medicine, encompassing the management of all human functions. An ethic of needs exists following the example of that of functions.

Moreover, the satisfaction of some secondary needs can be identified as a virtual necessity, where the cultivation of the ego generates functional derivatives in the same way as a drug addiction creates a dependence on an artificial world which creates new needs that bring an imbalance of the body.

Let us return to the metaphor of the human body compared to a sphere divided into four equal parts representative of our four major physiological functions. Each one corresponds to an extra-organic prosthesis, whether overgrown, degenerated or normal. Thus, we see a variable representation of these prostheses corresponding with each of the four functions.

Each prosthesis generates direct and indirect risks, which for man vary according to their development. For example, the risks linked with metabolic prostheses cause

16. CCNE: *Comité Consultatif National d'Ethique, Travaux du CCNE 20ième anniversaire* (Proceedings of the National Advisory Committee of Ethics), coordination Didier Sicard, Quadrige – Presse Universitaire de France, February 2003.

changes in the environment which interact extensively with physiological balance.

Indeed, nuclear power plants which provide, among other things electricity, increase production of radioactive waste, and thus increase the serious risk of human infection associated with storage and disposal. Without specific prevention, the body would suffer the effects of its environment.

By "high-risk society" we denote any society able to generate, by the application of its achievements, an individual and collective physiological functional imbalance, in the same manner as those of the autonomous regulatory systems of the body, which are unable to maintain homeostasis specific to the human species in its physical, psychological and behavioural characteristics. These are the societies likely to cause a homeostatic rupture.

This classification highlights the physiological and ecological limits of adaptation, which is the interactive and complementary expression of balance. Equally, it introduces the idea of genetic modification or epigenetics related to environmental changes.

Thus, our quadripartite society, as we have previously defined, can be interpreted as "high-risk society" because of its physical, chemical and biological capacities to irretrievably alter the physiological homeostasis, and also the genetic homeostasis, the impact of which cannot be measured.

On the other hand, a "low-risk society" lacks the capacity to alter the homeostasis to the point of jeopardising the human race. It is a tripartite society as described by Georges Dumézil in his analysis of ancient societies in India. This tripartition essentially translates the weakness of the metabolic sector (the production company) and, jointly, that of the group managing waste disposal which has so little influence that it does not obtain a representative social status. The metabolic sector is therefore taken care of by the neuropsychological and immunological groups. We can take the model of the evolution of French society. The French Revolution provides an example of the installation of a fourth social group. It is interesting to note that after the Revolution the French society slowly changed from a tripartite society to a quadripartite society.

Indeed, in the Age of Enlightenment, three groups were representatives of French society: First, the *Clergy* had a spiritual function emanating, on a physiological level, from the neuropsychological component. Divided between the secular clergy integrated into society and the regular clergy living in monasteries, it had honorary, tax and judicial privileges. Secondly, the *Nobility* had a military function and also the unique role of defending the kingdom against attacks. Its functionality was focused on protecting the organism, in the same way as the immune system physiologically defends against viral attacks. Finally, the *Third State*, which comprised of the middle class, the peasants and the servants. It was divided among the citizens who worked to produce and support the nation, playing a metabolic role by producing consumer goods; the peasants who provided food; and the servants of the social system who managed the lowest tasks, in particular the elimination of the waste produced by man. The industrial revolution gave rise to a fourth social group managing the country's industrial economy, individualising the metabolic component. It is pertinent to refer to the spherical model representative of the four major human functions. We realise that the social divisions depend on the development and management of major physiological functions of the organism.

Thus, the middle class managed the metabolic needs specifically, or in other words, the metabolic prostheses, by improving techniques and the production of food. It owes its growth to this improvement of production techniques and logistics. To achieve its purpose, it used the labour of workers. The middle class seized power over the nobility and the clergy and distinguished itself from them by giving itself quasi-exclusivity of metabolic needs. It reaped a financial profit, as it met and possibly generated needs for consumer goods. By means of the industrial civilisation, it individualised and appropriated the metabolic prostheses. So, the social divisions were between managers and agents. The metabolic systemic took shape in an infrastructure where middle class, financiers, economists, workmen, producers and consumers mixed. Thus, social quadripartition was formed. It really only developed on the waste disposal front during the 20th century, when its ever-increasing production brought with it a constantly increasing pollution – a major problem at the beginning of the 21st century.

The separation of the Judeo-Christian church and the state finalised this new functional approach to the society, grabbing from the church a part of its economic power. The neuropsychological systems then included, on the same basis, the representatives of all religions, the intellectuals and the philosophers. Through the logic of functionality, each group occupied a strategic and complementary place relative to the others. Thus, each social group became tougher and sought to impose its own representation, even if it meant establishing a dictatorship signifying its importance. The army, with its spirit of defence, kept its individuality in the same way as medicine defended health. The religious philosophers and the intellectuals came together or competed, giving rise to ideologies. Thus, from the mid 19th century, we see an evolution from the tripartite social system to a quadripartite system. Initially, we can distinguish those (humanists, intellectuals) who have knowledge and promote the emergence of ideas of those (religious philosophers) who are sources of ideologies. Moreover, economists and financiers are an entity managing the production tool. Then, the army is maintained and retains its individuality. And, in the last place, are the waste managers who gain more and more importance with the increase in industrial waste disposal. By developing his functionalities, man increases power and reveals the specific risks related to the existence of each social division.

Consequently, we sometimes see industry developing in the anarchy of private interests. If we take for example the sinking of waste ships sailing in defiance of international rules, the spread of avian flu in unhealthy farms, or mad cow disease where the type and source of feed are involved, the explosion inside a nuclear plant in Chernobyl irradiating thousands of people and condemning the land for hundreds of years, or the dumping of chemical waste into rivers and seas, it appears that man affirms his responsibility either by interest, ignorance, inconsistency, or presumption.

Examples proving the absence of human reasoning despite the capacities of the environment supporting these aggressions abound. Some think that the Earth sorts out these situations and that there is nothing to worry about. Supposing this is true, how long will it take? And, the Earth does not need man to survive; this is not the case for man in relation to the Earth. Adaptation, adaptation! Here is a word that will solve all the problems. But adaptation comes at what price? At the price of the non-expression of certain genes, the price of irreversible changes of the body, at the cost of the destruction of flora and fauna! Are we not slowly killing the planet Earth, for private or national interests, in burying one's head in the sand?

The risks associated with environmental changes interact significantly on the human physiological balance, which undergoes its effects and adapts to this context within the limits of its capacity. Are we not locking the regulatory mechanisms which enable us to live in balance with the environment? What are the limits of our adaptation?

Thus, every functional prosthesis is generating interactive risks which can become heightened. Today private ambitions are potentially amplified by these prostheses, which offer a new field in the approach of political responsibilities and capacities of power and of the destruction of minorities.

Controlled by an adaptation of laws and a permanent reflection on evolution, some countries develop more harmoniously than others, who possess overgrown or degenerated functional prostheses. Through their representation varying by country, each prosthesis generates direct and indirect risks which vary according to their development. Social inequalities appear within, even with the exposure to risk, prompting Ulrich Beck to say: "The proletariat of the global risk society lives at the foot of factory chimneys, beside refineries and chemical plants of the industrial centres of the Third World ... Extreme poverty and extreme risk attract each other systematically like magnets."[17]

With improved scientific knowledge, man begins to be studied on a biological level, gradually becoming aware of the power he holds on life. Using knowledge which he did not invent but simply discovered through research, he

---

17. Ulrich Beck, *La société du risque, sur la voie d'une autre modernité* (The Risk Society: Towards a New Modernity), Champs Flammarion, Paris, 2001, p. 75.

thinks he will become the master of his life. He thus creates a dreadful, pernicious global risk: that of locking man in a physiological prison designed by himself. This risk, which is not on the same level as the others, is the most terrifying. It is that of non-medical genetic engineering, where man, in his pretentiousness to impose his own point of view, wants to become the creator of life. So, he will "fiddle with" genes at the risk of creating monsters, centaurs! Does the dream become reality or are we reproducing a pattern which could be the cause of our decline?

How can we fail to think of the first page of *Genesis,* when man is walking in the Garden of Eden, between the tree of knowledge, whose fruits are not forbidden, and the tree of good and evil? Can we not see a parallel between the benefits of knowledge and the harm generated by pride and the power given to man, who wishes to become the master? If the satisfaction of our curiosity about the universe is legitimate, excluding its unthinking application by those who are not scientists, then it is in its most long-term and most responsible dimension.

"To manipulate life"[18] is to interfere with our origins and our future. It is to renounce a part of ourselves in order to design only what we imagine to conceive normalcy. It is in fact to consider that we can act according to our will and in our favour, to know all, to control the infinite, whether it be small or large. How can we fail to feel the ambiguity of some biological research which, in order to arrive at completely noble ends of curing genetic diseases, uses human material of embryonic origin? A dilemma for these researchers sometimes irresponsible! Is it morally right to use human embryos for experiments in order to seek

---

18. Ibid., p. 438.

appropriate treatments for currently incurable diseases? Some may say that the embryo not having speech is only human in his destiny, the cure of genetic diseases concerns human individuals. However, any experimental research on human embryos is a violation of the human potential represented by this embryo. Our ignorance of embryonic feelings allows some researchers to work on embryos that we torture and then murder without further ado. Does saving another suffering being warrant this silent crime? If nobody screams, everyone agrees! I have heard religious people speak about embryonic development and assert that, given the lack of sufficient cell differentiation, the embryo could not feel a murder. Yet, each cell of the embryo is marked with its own name (cloning) and has a potential of adult development. The embryo is individualised at conception, it is a fact. In the current state of knowledge, can researchers say that the human embryonic cell is undifferentiated?

Can we really change the laws of nature for the benefit of those who suffer injustices using the potential of an embryo? What might the long-term consequences be in patients carrying these embryonic transplants, and their offspring? Furthermore, advances in science now allow us to distinguish undifferentiated stem cells along the paraspinal region. These cells could, in the future, serve as a support for experiments and perhaps reconstruct some impaired organs. We must learn to refuse certain types of research using human embryos out of respect for the potential being that we manipulate.

Some of our acts thus put the environmental condition in competition with our outlook on life. The self-regulation of nature does not require our opinion and it responds according to its own criteria to all our ambitions and all our speculations. Yet who can claim to know the exact causes that lead to genetic diseases? We see that chromosomal damage was very difficult to explain until now.

We also talk about in vitro fertilisation: assuming there is a state of balance between the surrounding environment (the uterus) and the embryo, how do the genes react during an in vitro fertilisation? What influence does the changing environment have on the human genome? Might it not cause the impairment or a non-expression of some genes which could be transmitted from generation to generation? Even if I understand that this issue is disturbing because of the daily practice of in vitro fertilisation, the question still remains relevant. Moreover, among the possible drifts, this operation could make us lose sight of the natural complementarity between men and women by 'mechanising' fertilisation. In fact, most couples using in vitro fertilisation are suffering because of their difficulty conceiving. It is necessary, however, to ask this question and hope that the researchers of the future can show that in vitro fertilisation does not disturb genetic homeostasis.

Approach to Functional Balance in Indian and Chinese Societies

Evolution of the Social Divisions in India

In the preceding chapter of the book, we explained how a physiological reasoning applied to a high-risk society was able to facilitate modelling of the management of each functional component. We have, according to our demonstration, put forward the hypothesis that high-risk societies develop social quadripartition. Yet, as Madeleine Biardeau maintains in the introduction to Mahabharata[1], the Indian social model was primitively based on a quadripartite management of society. (To facilitate the comprehension of

1. *Le Mahabharata 1*, Introduction by Madeleine Biardeau, Flammarion, p. 24.

the text, certain words of Indian origin have been specified in a glossary at the end of this work.)

"The theoretical quadripartition of Brahman society forms part of its mental framework (much more surely than the tripartition of Dumézil)."[2]

The Mahabharata is a reference text for approaching social quadripartition and its subsequent tripartition. It transcribes an oral tradition, mixing history and mythology where it is sometimes hard to tell truth from legend. It conveys a message which one must know how to interpret. Above all, it transmits the memory of an event provoked by human behaviour, whose consequences were catastrophic for the humanity concerned, analysing the causes and offering solutions. Thus, considered sufficiently important and urgent, the revelation that this text attempts to pass on from generation to generation has come to us through a social system showing it to be a remedy to the folly of mankind.

The story goes back so far that the original event which was responsible for this state of mind is not known:

"We know nothing about this first major event that forms the basis of this collection (The Mahabharata) as an intangible revelation."[3]

The detailed memory of the event would have been lost during the course of the centuries, whilst leaving an indelible print on the history of mankind.

"It is not forbidden to think that, when the notion of revelation was developed, there was no longer access to a complete understanding of acknowledged founding texts."[4]

---

2. Ibid., p. 24.
3. Ibid., p. 25.
4. Ibid.

This text consists of a social response to an event for which mankind appears to be responsible and for which it feels guilty.

The distribution and composition of Indian social groups are established and structured throughout this text. Its normative structure, bringing together the management of all the important human functions, gives a universal and timeless behavioural answer to functional human needs.

Indians "think of themselves as living under a sign of permanence. So we must be careful: they exist as if they were living in a fixed universe, with perfectly stable references, with uncontested values".[5]

The distribution of Indian society owes nothing to chance. It is a given profound significance that has to be interpreted as coming from a revelation. It is the concrete expression of memory. However, although numerous deviations or drifts have appeared during the course of the centuries, it is nevertheless true that the foundations of the social system persisted.

"The political fragility revealed in the writings is integrated into this system, where the profound stability of norms rules."[6]

We have previously shown the evolution of French society from a tripartite system to a quadripartite system, the latter corresponding to the birth of industrialised society, which in the following centuries, and even more so today, would generate a high-risk civilisation/society. Yet, since its foundation, the Indian social system has been quadripartite, and therefore adapted to such a society. Whilst this social functional classification did not seem to have been suitable

---

5. Ibid., p. 18.
6. Ibid.

for the needs of the society, it was nonetheless imposed by relying on a theoretical argument which did not give rise to fundamental criticisms of these principles.

For such a system to be established, it was necessary to prove its relevance, to provide sufficiently convincing reasons for them to be accepted by the population, to respond to a feeling of urgency in a particularly worrying context, where feelings of guilt and fear inspired by human behaviour dominated, and to judge that an event provoked by mankind was sufficiently serious to lead to major social reform.

Several points attract our attention, for example, the social fabric of the system has not changed for thousands of years. If the system did not correspond to the management of physiological needs, if it did not exactly relate their functionalities to their prostheses, it would not have lasted. But, as Madeleine Biardeau points out: "The permanence of the frame of reference does not exclude all change."

Indeed, this quadripartite society has evolved towards a tripartite society over the course of centuries. It is ultimately the application of the quadripartite system to a low-risk society that poses a problem and leads to the search for an explanation of its social inadequacy.

Having established such a system in a low-risk society proves that the theoreticians who set it up had experience of human behaviour beyond that of the society they lived in. Most likely they had analysed the causes of a disaster engendered by man. To have such a system be accepted by a population proves that the theoreticians' message was heard and understood, and that the arguments used in favour of this type of social organisation were sufficiently convincing not to be questioned at the time.

This system seems to have been established with a feeling of timeless urgency, much more focused on human behaviour in general than the will to act in the short term. This means that the theoretician(s) who established this social system, conscious of the message that he (they) had to transmit, knew that it was not adapted to the period to which he (they) had applied it. Perhaps it would be, one day in the distant future. He (they) had a vision of human potential in its capacity to destroy itself, but also in its capacity to protect itself and take responsibility. It was an act of prevention in relation to risks that had appeared in the past, or could occur in the future. This system evokes for me the painting by Mendeleïev where boxes are progressively filled as new elements are discovered!

In this context of anxiety about the future, normally a duty of memory is imposed, which results in the endurance of the established social system. Rather than ask, "What happened?" the descendants preferred to ask themselves, "How should I act in the future?" and they reacted on the behavioural level according to the past event. We will analyse this behaviour in the study of antithetical memory.

Every normative system sets up a defence, a response to a pattern of disorder whose causes must be sought. Every normative system can only last if it is representative of an intangible reality. It is here that physiology intervenes. To live, we will always have four functional interacting components that can develop capacities by prostheses. Furthermore, we know that there is a homeostasis regulating the organism – a system both dynamic and adaptive that permits us to live in a variable environment. In addition, the Indian genius was to impose a jurisdictional system between the social groups in such a way that the management of needs is interactively regulated. It is an unstoppable system of prevention that relies on the inertia of the ensemble. Certain people rightly maintain that this system blocks initiative and evolution, meaning, in their words, "progress". This would be true if human, scientific and technical knowledge were only presented as positive aspects and did not involve the risks of upsetting the homeostasis, eventually leading to a human disaster. This point is fundamental, because it brings into play a possible vision of the application of human discoveries. One would think that the theoreticians who put this system in place were perfectly aware of this. It is well worth noting that they did not seek to apply a sanction, but rather behavioural regulation in the face of dangerous psychological risk-taking.

All the historical events that would follow corresponded to the normal evolution of this society, where the omission of primitive behavioural logic would erase a part of the rigorously transmitted message. The substance would remain nevertheless, but the basis would change. It is under these conditions that social evolution would move towards a tripartition, due to the weak social representation of the group managing eliminations. In fact, the metabolic prostheses of low-risk society would generate only very little toxic waste, whilst our contemporary societies produce, in large quantities, toxic waste of physical, chemical and biological origins.

The quadripartition of the Indian social system, coming from the revelation, constituted for every society at low risk a relatively illogical or, at least, ill-adapted innovation. Without necessarily being compelled to, it would evolve little by little towards an actual tripartition, conforming more to the reality of society today. It is by the principle of the division of the tripartite social structure and the type of ideology which followed (realised or hoped for) that Georges Dumézil demonstrates the importance of these functions. Thus, he opens the way to our preceding reasoning: "Many times, (...) an important word was encountered: that of functions, the 'three functions'. Certainly, it should be understood by that, the three fundamental activities that men – priests, warriors, producers – should guarantee in order for the community to subsist and prosper."[7]

Within the Indian social system there is an ideal of justice. Throughout our analysis of the use of prostheses, we have noticed that some of them were subject to hypertrophy and others to hypotrophy. This system engenders disequilibrium between groups managing these functions and ultimately social injustices responsible for disorder. Each person must do his work as a function of the group to which he/she belongs in order to limit excesses. The jurisdictional system that has been translated under the terms of 'socio-cosmic order' (Dharma) deals with the practically legal interaction between the social groups. It maintains homeostasis, thus avoiding disorder.

It is quite possibly by caricaturising that the system drifted towards an even more normative compartmentalisation having given birth to the caste system. If the physiological make-up remained, the contents of the social groups adapted.

Reflecting on the concerns of such a system, it would appear that its quadripartition facilitates the management of four big risks: those linked to production, ideologies, arms and waste.

If this is applied it to contemporary Western society, we notice that the galloping increase in waste production brings with it the crucial problem of its elimination, given the ecological and physiological disequilibrium that it can engender. The representation of a social group charged with

---

7. Georges Dumézil, *Mythes et Dieux des Indo-européen* (Myths and Gods of the Indo-European) Flammarion, p. 95, 92.

the administration of its management is now accepted and has become customary. This awareness must have been much less clear for the ancient Indians during the creation of the quadripartite social system. In fact, the representatives of this social group only regulated problems of health and hygiene, without having to be responsible for waste, as we know it now. Their presence was undoubtedly for a reason that escapes us.

Indeed, by evolving over the centuries, the Indian social system did not completely reject quadripartite, but dismissed its function as not being sufficiently socially representative.

It must bc admitted that the quadripartition of the Indian social system is directly related to an event whose trace has been lost and was established in order to avoid the same type of event recurring. Indian society is built on a fixed universe, with uncontested values. Indians express themselves as if they already knew the basis of the problem. As long as the social order is respected, what does political power matter! We have, then, dealt with the transmission of the Vedic Revelation based on wisdom and where the standard prevails.

Since the works of Claude Bernard, physiology has helped us to better understand the role of the functional components and regulatory mechanisms of the body. The 'steady state' of W. Canon, defining homeostasis, is close to the vision of stability that Madeleine Biardeau talks about. Stability and regulation are two fundamental concepts in Brahman society. Indira Gandhi writes in a book devoted to the India of today: "The concept of unity is almost as old as India itself"[8] and also adds:

---

8. Indira Gandhi and Jean-Louis Nou, *Inde, hommes, rites et dieux* (India, Man, Rites and Gods), Edita, 1978, p. 7.

> No Indian is exempt from the mix: he can be of the Dravidian, Aryan, Australoid or Mongolian type; his hair can be light or dark, straight or curly; his skin can be any shade from white to black. Because India has always accepted all the ways of living and thinking of the races and tribes that people her, without demanding a conformity which would deny their individuality, however it marks them with the indelible print of their "Indianness".
>
> Ideas about life, the goal, spiritual and aesthetic aspirations bring this apparently diverse nation together as an integrated nation.[9]

It is possible to imagine that physiology permitted this coming together by providing the Indian social system with a stable, timeless foundation, an indispensible structure in order to explain the persistence of the same social fabric, whose sole objective is to maintain a state of equilibrium favourable to reproduction, life and love over the course of the centuries.

Physiological Functions and Jurisdictional System of the Dharma[10]

The two physiological functions of the body, most representative of individual identity, are the neuropsychological and immune systems. They are considered to be the most noble by Indians.

The metabolic component possesses a useful energetic function that ensures mobility. It is based on diet, the

9. Ibid., p. 7
10. Dharma is defined as "the ethics of life" (Indira Gandhi). It establishes the rules of life between social groups. It is interpreted as the universal socio-cosmic order. Indeed, it corresponds to a legal system.

production of energy products and their use, and finally, on economy.

The management of eliminatory products affects hygiene and interferes with the environment on a global level. It does not correspond, in itself, to a noble act if it is reduced to social necessities (hygiene). It becomes noble when it conditions the environmental equilibrium and, by interaction, the human physiological equilibrium. It becomes important in all societies at high risk.

In ancient Hinduism, society was divided into four Varnas: the Brahmans, linked to the sacred, were the priests, teachers and professors. The Ksatriyas, possessing temporal power, included the kings, princes, administrators and soldiers. The Vaishyas, linked to the clan (of production), comprised artisans, sellers, businessmen, farmers and shepherds. Finally, the Sudras were the servants.

A central theme of the ancient Indian social system would seem to be the physiological equilibrium. Its objective is the balanced management of human behaviour through its functionalities. As we have suggested, this approach basically necessitates a precise knowledge of the functional components of the organism as well as the knowledge of how they interact. The joining of functional extraorganic prostheses, as we have defined them, allows us to understand how a social organisation can be developed wherein the representatives of each group exert a level of responsibility according to the function they assume by corresponding with the other groups. This normative principle can be disrespected and provoke a social disequilibrium (Adharma) bringing a mix of Varnas.

"Any combination of Varnas can only cause trouble or be a sign of social disorder."[11]

---

11. *The Mahabharata*, op. cit., p. 33.

It should be noted that the quadripartite management of Indian society decentralises the power that is thus situated at the level of each group, then between the groups, due to an interactive jurisdictional system described as a socio-cosmic order and called 'Dharma'.

How can we not identify this functional approach with the vision of Ulrich Beck who remarks that the principles of flexibility are "increasingly preferred in situations of risk and insecurity, but also with new forms of auto-coordination of the subsystems and decentralised units of activity."[12]

His vision of social regulation that operates interactively between "decentralised units of activity" strangely makes us think of the Dharmic organisation of Indian social groups.

Power is not centralised, but the Brahman function (neuropsychic function) monopolises through its knowledge of the Vedas (Book of Wisdom), a philosophical and religious influence, and benefits the Varnas by "its monopoly over rituals and all normative knowledge; it remains, due to its secret knowledge of the socio-religious order."[13]

We need not be surprised that each social group plays a specific role, while depending on other social groups within the framework of a hierarchy linked to knowledge, defence, production and elimination of waste. The metaphor of the functioning of the human body is frequently used to explain the organisation of social groups into a hierarchy.

It should be noted that every individual possesses autonomous and independent physiological functions. The division of the Indian social system, as well as the specificity of the roles allotted and achieved by each group, lead us to think that the theorists of the Varnas had a very good

---

12. Ulrich Beck, op.cit., p. 486.
13. *The Mahabharata*, op.cit.

knowledge of human physiology. The particular role of the Ksatriyas, whose way of life consists of compulsory violent exercise, both in extreme defensive or offensive roles, gives a sacrificial image of this group, which has persisted throughout the centuries. Nonetheless, it would seem difficult to understand the reasons that lead Indians to associate the defence of the Dharma with the sacrifice of the Ksatriyas.

Is it necessary to see in this excessive role of the warrior who sacrifices himself for a social balance in respect of Dharma (the socio-cosmic order) an unconscious analogy with the action of the immune system of the body? Same excess, same extremism, same violence, same outcome! It is certain that the Ksatriyas struggle violently for an equilibrium in society as much as the immune system struggles for individual homeostasis.

"The Ksatriya must make war, be it against his brothers, when the Dharma is at stake, and make this war, despite its repugnance, in order to come to the rescue of the Dharma, for 'the cohesion of the worlds' which would, without this, return to chaos."[14]

It seems to me difficult to conceive that this quadripartite social system would be able to be established without being inspired by experienced theoretical knowledge. Why does the Ksatriya have such a normative obligation to action? What are the reasons that lead the theoreticians of the Indian social system to adopt such rigour in the defence of the society? Isn't there an underlying fear of seeing the emergence of a social imbalance (Adharma) related to the disorder which can generate a disaster? Will history eventually play a role?

---

14. *The Mahabharata*, op.cit., p. 37.

If the management of hygiene proves indispensable in tripartite societies (at low risk), it is however difficult to imagine that it can unify around this theme an ensemble of representative individuals of a sufficiently federated group to assert itself in Indian ancient society. Its rejection establishes greater proof of the inadequacy of the theoretical social system during its creation with the society at the time. The passage from a quadripartite to a tripartition would appear as a fundamental proof of the perhaps wilful inadequacy of the system in place with the reality of the moment. Why set up a system which is inadequate for the needs of the era? Was the quadripartition not promoted in response to ancient events by duty of memory?

Chinese Bipartition and Indian Quadripartition: The Search for Social Equilibrium

The Chinese have a perception of balance as obsessive as the Indians. It seems to me difficult to accept that it is mere coincidence. These two nations were literally saturated with a cult of equilibrium showing us their inexpressible fear of human behaviour. Each of these civilisations searched for laws of balance, and for thousands of years spread the complementary vision of prevention.

However, they differentiated themselves by their visions of balance. If the Indians seemed to have relied on physiological observation of the human organism (homeostasis), the precursors of the Chinese system of thinking have a much more physical approach to balance, underlining in a dualist and paradoxically complementary way, the matter-energy equivalence. In both cases, the human organism is at the heart of the debate, each people giving their recipe for keeping it in a state of equilibrium. Whether it was Tao law transmitted by Lao-Tsu or the socio-cosmic order represented by the Dharma, rigorous rules would regulate human behaviour. Thus, the Dharma established legal rules to be respected between social groups. And, for its part, the respect of Yin and Yang would be imposed on Chinese society.

If one analyses the Chinese message as a concept, he/she could consider that it expresses an Einstein-like vision. The equation $E=MC^2$, the mathematical expression for energy (E) and mass (M) would appear as a more concrete and generalised projection, like a "matter-energy" equivalent. From then on, one question seems to be fundamental: why have we envisaged and created a system of thinking, which undeniably inhibits behaviours by seeking to balance them? The specificity of Yin-Yang actually creates a schism that can lead to a form of dualism. This opposition is also paradoxical since Yin and Yang are considered as two complementary expressions of life. The junction or very narrow point of fusion, which could resemble an organised, empty path, separates the Yin from the Yang in Taoism, making it difficult to interpret the Taoist texts if this concept is not accepted.

The visible and the invisible seem to be two different entities in the same way as matter and energy. In fact, matter is visible and energy is not. Yet it is a question of the same reality, relying on the same mode of reasoning, expressing itself through two independent and complementary states, evoking the book of mutations.

"They are the same (and unique) reality understood at two different moments. From that moment on, the visible and the invisible turn out to be independent."[15]

---

15. François Jullien, *Figures de l'immanence, Pour une lecture philosophique du Yi King* (Figures of Immanency: Concerning a Philosophical Reading of the Yijing), Éditions Grasset et Fasquelle, 1993, p. 233.

It should also be added that these complementarities possess a dynamic, which imprints a form of balance onto existence: "These antagonisms do not oppose each other implacably; on the contrary, they imprint a balance on things which rather more aim to guarantee equilibrium."[16]

Our visual apparatus, thanks to technology, permits us to obtain vision of structures invisible to the naked eye. It is thus possible to push back the boundaries of the invisible. The visible includes a part of the invisible that one can make artificially visible. Energy, as it is defined, can thus be made visible, further pushing back the boundaries of the visible.

Today medicine can penetrate the invisible world of illness, thanks to numerous recent techniques – be it physical (microscopy, scanners, MRI, etc.), biological, physiological etc., examinations. This is what causes Michel Foucault to say in *Birth of the Clinic*: "The abyss beneath the evil, and which was evil itself, is exposed in the light of language."[17]

As Einstein has shown, a simple relationship between energy and matter exists. How is the balance established between matter and energy? Can one delineate a zone? Does it move?

Indeed, a state exists beyond which one cannot go, since one can no longer see anything, and since the visible and invisible become confused in a sort of vacuum, giving way to all that is unsupported. Beyond energy and matter is the void, which has no expression or reality in the material sense, but whose existence, systematised by

---

16. Jean Grenier, L'esprit du Tao (The Spirit of the Tao), Champs Flammarion, 1973, p. 48.

17. Michel Foucault, *Naissance de la Clinique* (Birth of the Clinic), Presse Universitaire de 7ème France, 1963, reprint of the edition "Quadrage": February, 2007, p. 199.

mathematicians, could one day be represented by investigative techniques more modern than the scanner or IRM, analogous with a constant in the human organism perhaps like the Chinese meridians used in acupuncture:

"Thirty rays come together around a hub. It is on its emptiness that the use of the tank depends.

We knead clay in order to make a vase. It is on its emptiness that the vase's use depends.

We make doors and windows when building a house. It is on their emptiness that the house's use depends."[18]

The human body would have been constructed by linking matter and energy in a symbiotic way, and by leaving an empty space as a free zone.

Would the realm of the conscience not represent the existing interaction in this indefinable zone, beyond the visible and the invisible? Does it not correspond to a state, a personal measure, and a space for judgement allowing one to see oneself and to see without entering the matter-energy dualism?

Can the permanence of a true observation of oneself, combined with the impermanence of our material and energy state explain the conscience as a state on which we have no hold but an influence? What persists is the void, at the frontier of matter and energy, where an objective look at ourselves and others gives a sense of eternity. And thus we wait for the technical progress that places us at the margins of the metaphysical.

"The ephemeral visible is only an appearance, opposed to the truth of the eternal", wrote François Julien in *Figures*

18. Jean Grenier, *L'Esprit du Tao* (The Spirit of Tao), Flammarion, 1973, p. 16.

de l'Immanence (Figures of Immanence). It can be translated thus: The reality which goes beyond the visible is a part of the void filled by an indefinable equilibrium which leaves aside matter and energy. This space, difficult to conceive of at the moment, will perhaps be found one day by scientific methods.

The Chinese and Indian ways of thinking rely on facts that refer to both transmitted knowledge and an acquired historic experience.

We can try to find a part of the story by calling on what I define as 'antithetical memory'. We begin with the principle that every period of disorder and disaster provoked by man can be analysed and memorised by those who survive it. A regulatory reaction follows, whose goal is to avoid that catastrophic events linked to such social disorder are not repeated. If one attributes the causes of these disasters to human behaviour, then every social system seeks to avoid its repetition by struggling against social deviations. The thesis that leads to the disaster is succeeded by the antithesis, which leads to wisdom.

Human Behaviour and Antithetical Memory

History is based not only on facts, but also on reactions. Facts nourish the imagination and the memory, amalgamating the real to the virtual. In addition, fear, guilt and other reactive feelings can result from quite precise reasons which become lost over the centuries. It is also true that the collective memory records the reaction as a fact that it will be able to transmit throughout the centuries. A reactive message can lose trace of the reason for its origin. It does not transmit any less of a revelation, which will have to be interpreted, in the same way as an event whose cause would not have been known at the moment it occurred. The fidelity of the transmission of the message constitutes the best weapon against forgetting. The historical reasons for forgetting are interesting to analyse. There is the wilfully forgotten – either the burden is too heavy to bear, or the desire to be free from it is stronger than the fact of having been subjected to a story that is ignored and which inhibits behaviour. Forgetting can provide the intoxication of the present moment, the enthusiasm derived from the improvement of the standard of living, by a certain feeling of existing and being respected. Memory can be upsetting and the reasons for forgetting can appear superior to the fact of remembering and restraining oneself. Wilfully forgetting can thus permit the separation with the past by favouring the reasons for forgetting. However, "The forgotten remains the unsettling threat which proliferates in the background of the phenomenology of the memory and the epistemology of the history."[19]

That is why, the ancient Indians, conscious of the importance of the message that they had to transmit, applied themselves to exercise the memory of their descendants by imposing a very strong oral tradition allowing the struggle against forgetting.

As has been previously mentioned, India founded its social philosophy on the management of physiological balance. This fact is not insignificant in itself, and we should ask ourselves about the maturity of mind of those who had this conception. This intangible data, being related to the management of physiological needs, has persisted in the population for thousands of years. India was constrained to experience, because of its functional approach to

---

19. Paul Ricoeur, *La Mémoire, L'Histoire,* L'oubli (Memory, History and Forgetfulness) Editions Seuil, 2000, p. 536

equilibrium, a normative model of behavioural wisdom, even during political changes. A revelation whose origin is effectively ignored constitutes the basis of this civilisation and remains anchored in people's mind as a *reason for being*. By the fruits of their reflection on life, on human behaviour, the Indians have long resisted the assaults of Western economy, placing themselves outside the global economic system, thus remaining a behavioural mystery for many of the players in the so-called modern nations. Devotion, misery, the need to redeem oneself and also to make sacrifices presents the image of a timeless India, carrying all the burdens of human errors. If the current model of the castes scandalises through the inequalities it creates, it should be seen as the result of shift that took place over the course of centuries. Indeed, the object of this study is not to make a judgement about what is happening currently in India, but what took place thousands of years ago and which has caused enduring behaviour that our Western logic seems to have difficulty in comprehending. The study of behavioural reactions in a pre-factual society can give us a better understanding of their earliest, ancient behaviour.

If, on the one hand, one considers the evolution of Indian society using Western logic, it seems difficult to understand this behavioural amalgam. On the other hand, if one moves away from the hypothesis that this society was born of the imagination of one or more theoreticians having acquired great experience of human behaviour in societies at high risk, it seems to me that one better understands the reasons that lead this civilisation to adopt the functional division of the society. We are not speaking about socio-biology but rather of the management of physiological functions that we are all endowed with.

In fact, what interest can be found in dividing the society into four big groups? Firstly, these groups have to be representative. Let us say that the groups of the Vaishyas (metabolic function), the Brahmans (neuropsychic function) and the Ksatriyas (immune function) occupy a predominant place in a society at low risk. On the other hand, the group of Sudras (eliminatory function) is concerned with waste, and only really becomes important in societies at high risk.

Returning to what I call extra-organic functional prostheses, we have seen that every high-risk society acquires technical capacities for manufacture which themselves generate waste that needs to be treated. For example, in The Hague, the nuclear industry stocks and treats radioactive residue. If one adds all the waste which has to be treated, we realise how specialised the work is, as well as the social representation of the individuals charged with its management. From then on it is conceivable to bring together an ensemble of individuals, capable of assuming responsibility for the elimination of this waste, around a common objective.

Imagine that one or some theoreticians, having survived the event provoked by the human unconscious, looking for a social system in relation to high risk. Given our current knowledge of the secondary effects of waste elimination, it would appear justified that a group of individuals specialised in this management has social representation. It is this hypothesis that gives the Sudras reason to exist. However, if a catastrophic event provoked by man destroyed all waste-creating industries, their management would be considerably reduced. The group of Vaishyas, that is those who manage the means of production, would no longer have the capacity to eliminate toxic products. It would seem natural that the social representation of the Sudras loses a large part of its profound reason for being. The forgetting of their function and their social representation would become blurred over the course of the centuries. So they would lose a large part of their influence, since they would no longer have sufficient representation by comparison with the Brahmans, the Ksatriyas and even the Vaishyas who would keep their nutritive production role. They would be condemned to deal with the lower tasks, in particular everything that is concerned with hygiene, leading a certain number of them to be rejected as untouchables. If the Indian social system endured through so many centuries, it is because of the intelligence of its functional representation and its mode of regulation (Dharma). It is conceivable that this was put in place in order to avoid human deviations, and to avoid the recurrence of an event, a part of whose history the Indians carry, a burden which has generated a considerable and persistent work of memory due to the responsibility which lies with man who must morally transmit the memory of what he has provoked. India, through its social system, warns us of danger. It is for us to understand! One can only be appreciative, faced with such efforts of memory in order to transmit a message, a very clear warning, if he/she has the humility of the interpreter and the logic of our current knowledge. Does an 'archaeology of memory' which transmits the real experience of a situation from generation to generation without any writing exist?

This reflection leads us to ask ourselves the following question: Were the theoreticians who put in place the social system of the four Varnas conscious of the inappropriateness of their system for the society at low risk in which they were living or, with a long-term vision of the future, did they solidify the system by normalising it in order that it survive throughout the centuries?

According to the previous reasoning, this system represents a universal aspect because of the functional normative expression of each social group. According to its regulatory volition, it has certainly slowed down the Indian economic dynamic over the centuries.

"Besides which the socio-religious context did not compel scientific research beyond immediate requirements, the conception of the human mind neither permitted great theoretical daring nor renewal of perspectives."[20]

Whilst the economic and industrial shock is currently occurring in India, and Indians admire the rapid development of industrial societies (through rose-coloured glasses), they are conscious of the inhibiting effect of the castes on their economic evolution. If the ancient Indian social system modified itself over the centuries in order to lose part of its efficiency and regulatory mechanisms, it must be admitted that it had naturally drifted for thousands of years, establishing power relationships where there was no reason for being. However, and this is the genius of the system, due to the fact that it is connected to human physiology, the social fabric has endured unto this day. Today, its ancient application, liberated from its origins, possibly allows the effective struggle against the risks of our industrial societies. According to this thesis, would the Vedic Revelation and the establishment of four Indian social groups not be destined for us?

Let us admire these Indians who carried a burden throughout the centuries. Let us understand their need for liberation, and above all, let us interpret the message that they transmit to us, which causes us to evolve our management of risk.

---

20. Madeleine Biardeau, *L'Hindouisme, Anthropologie d'une Civilisation* (Hinduism: Anthropology of a Civilisation) Flammarion, 1995, p. 255.

Could this behavioural evolution, of the Indians and Chinese alike, provide a partial explanation for the apparent brutal change in their behaviour? Is it not necessary to forget in order to acquire new benefits, knowledge, dreams and to live in another way? Will their philosophy impede our development today? Does the philosophy in itself not constitute an interpretation of events, and is it sufficient to replace experience? Should the elders and their messages on equilibrium be believed? Should philosophy lead to poverty or poverty lead to philosophy?

It is in the study of interpretations that antithetical memory, raised in a previous chapter, comes in.

We define by 'antithetical memory' every psychic and behavioural reaction provoked by an event, memorised and transmitted with the aim of modifying subsequent behaviour, in order to prevent the recurrence of a similar event. It is possible to retrace, through these memorised and transmitted reactions, an anterior history (the thesis).

Antithetical memory is to history what treatment is to illness. Knowledge of a therapy can enable the identification of an illness.

By these very marked behaviours and the clarity of the themes that have been tackled, Indian society offers us a particularly rich example of antithetical memory. The whole system is based on the adaptation of behaviour to a message, a revelation, so as not to repeat the same mistake.

The gravity of a situation, and its imminence, demands a feeling of urgency.

If one accepts the temporal vision of the Indians, he/she notices that their perception of time is very different from ours, and its very cosmic approach causes them to rapidly lose the notion of months and years. Indeed, the totality of the divisions of time "would be intelligible if it was not related to the idea of the creation of the world and it's perpetually repeating re-absorption."[21] We are living between the beginning and the end of time, and in a perpetual manner. We should know this sequence in order to prepare for it. We are not situated within a human logic but in a cosmic logic of events. Man undergoes these phases, which translate on earth as order and disorder – Dharma and Adharma.

Consider the hypothesis of a catastrophic event which some individuals survive. The devastation of the whole of civilisation could not only lead them to a reaction of despair faced with such human irresponsibility, but also to a need for hope constructed uniquely on an organisation of the future expected to protect mankind from its own behavioural madness. One can put forward the hypothesis that for millennia, the inhabitants of India have maintained their memory by duty, so as to favour the transmission of information in order to warn us of an imminent danger caused by man. They made sacrifices (and perhaps still make sacrifices) in order to bring us a message that we should now decode. This idea of duty and sacrifice is present throughout their history. However, it will be interesting to follow the current evolution of Indian society.

The immense work of memory accomplished in India constitutes in itself a major test of the feeling of a population that has been asked to urgently transmit a message, a revelation.

But why such urgency?

For them, history has already been written: time is structured in Kalpa[22] formed by a thousand times four

21. Ibid., p. 164.
22. The Kalpa is the name of the day of Brahma, between two cosmic contractions of equal length, which are its nights.

Yugas[23] or "great Yugas", being twelve thousand years from the gods. One year of the gods is worth three hundred and sixty human years. Their overall vision narrows the notion of day and night. And so, according to the Indian calendar, we are currently experiencing the worst age, that of the Kaliyuga. This fact can partly explain the urgency of the message.

"The Hindus agree with all the peoples of the world by saying that we are living in the worst age, one where social disorder is at its height: an idea which explains the gap between norm and reality."[24]

The absence of traditional recording techniques has encouraged oral tradition, the only means of transmission under extreme conditions. This could only astonish Madeleine Biardeau who, speaking about the Mahabharata, is surprised at the extreme rigour of the techniques of memorisation: "Had they not spent the very early morning hours of their day carefully memorising these texts according to the methods which today leave us flabbergasted, methods which assured the faithful transmission of the form of the discourse more than its comprehension?"[25]

Let us analyse the themes emphasised by antithetical memory.

Let us first speak about the Dharma or socio-cosmic order. Following a social disorder, which transformed itself into a disaster of great amplitude, the conception and the

---

23. The Yuga is the name given to cosmic periods grouped into four of which each one indicates a given state of perfection or degradation of the Dharma. The avatara of Vishnu generally appears with the passage of one Yuga to another.
24. Madeleine Biardeau, op. cit.
25. *Le Mahabharata* 1, op. cit., p. 26.

establishment of a quadripartite system of the Varna permitted the recreation of a social order.

It was necessary to envisage a system likely to contain the disorder. What was this disorder about?

Most likely a social disorder is provoked by an inadequacy between the management of needs and human functions. We should not lose sight of the fact that the human organism is an autonomously homeostatic system, which means adaptive and self-regulating. It possesses four main interactive functions that we have talked about previously, and which can be separated conceptually. A social disorder can stem from, for example, bad management of everything that concerns metabolism, which means from the expression of needs to the exercise of its functions through the formation of specific extra-organic prostheses or EOPs (train, plane etc.) that increase the mobility of the field of action. A hypertrophy, in the same way as a hypotrophy of the prosthetic functions, can be linked to the economy, finance and production, generating a major disequilibrium of the entire social system.

Why group each human function in such a systematic way?

In order to stimulate responsible behaviour, the normative side of the social system certainly gives us the key that allows us to analyse part of the causes that lead man to create disorder. To form a group which can be individualised, it must have sufficient social representation, meaning that it is recognised as having a specific function.

Our westernised society provides an example of the division of specialities. Today, the specialisation of functions, caused by the addition of functional prostheses, facilitates the regrouping of our vital forces by social groups, whether it be economists, soldiers, intellectuals or actors in the ecosystem. We have specific representatives from the army, the church, the economy, ecology, industry, medicine, etc.

However, unlike the Indians, our legal and political model is not completely adapted to a new sharing of functions. The Indian legal system, based on the Dharma, generalises the relationships between individual social groups according to their functionality. Our political power is indeed less and less adapted to the management of risks provoked by major functional centres, hence its weakening. Furthermore, a specific recognition of the legal framework allowing management of the interactions between these groups does not exist. An inter-group legal and political representation would favour exchanges and play a regulatory role. It would permit experts to come together in order to manage risks and synthesise balanced solutions for the nation.

A second and no less important point concerns the establishment of four Varnas and not three. Why establish a social group managing eliminations? Firstly, it seems excessive in a society at low risk. On the other hand, in a society at high risk, the management of elimination brings, besides those which are natural to man himself (hygiene), those which require a specific technology, for example in chemistry, biology and physics, like that required for radio-activity, which we have discussed previously. The existence of antithetical memory undoubtedly allows greater under-standing. Only a high-risk society can engender four major, specific risks, including those provoked by eliminations. That would indicate that the society which preceded it had attained a level justifying a quadripartite social represen-tation. Following a catastrophic event provoked by man, the survivors set up a protective system that takes into account their perception of the risk. The theoreticians, who knew of the event and, without doubt its causes, had conceived a social system adapted to a society at high risk, so that it had evolved towards a society at low risk. Through its pragmatic side, the latter casts aside the weakest, the least representative, and thus alters the original theoretical conception of the founders of the system. However, the social fabric of the system subsists of and remains anchored to the population throughout the millennia, until now, is not called into question. This approach creates no opposition between Madeleine Biardeau who asserts that the system of the Varnas is originally made up of four groups, and Georges Dumézil who analyses ancient societies on the tripartite level.

A third point tackles the obsessive behaviour of the Indians centred on the question of balance. Can a link be seen here with Chinese preoccupations that gave rise to the concept of Yin and Yang? Differing from the Chinese who started with a physical principle of the equilibrium existing between energy and matter, the Indians relied on a functional approach to human physiology in order to establish a social system. At the heart of every society are man and his capacities for adaptation. Due to the immense progress in medicine since the beginning of the 19th century, we are now capable of dividing the main physiological functions by taking into account homeostasis, as conceived by Jack Baillet and Erik Nortier in their *Précis de physiologie.*[26] So it must be concluded that, according to an antithetical vision, the society having preceded 'the catastrophic event' was profoundly destabilised and incapable of promoting the maintenance of social harmony. In addition, it knew the dangers of an energy-matter disequilibrium (nuclear energy?), and conveyed the necessity of

---

26. Jack Bailet and Erik Nortier, *Précis de Physiologie, Elipses,* (Handbook of Physiology, Elipsis) Editions Marketing, 1992.

respecting the Yin-Yang equilibrium. Is it the same event that determined ancient Indian and Chinese behaviour?

The fourth point refers to guilt. This is because everyone did not act as they should have and avoided their responsibilities, like Bhisma[27] who renounced marriage and refused royalty because of his father's wife. Disorder was born, leading to an absence of landmarks, to error and the forgetting of the other, to egotism and, in the end, to disaster. The entire history of India is impregnated with repenting that which has been provoked (and ignored), generating sacrifices (Bhakti) in order that the wrongful acts of its ancestors are pardoned. We meet the biblical notion of "original Sin".

If man is not held to rules of behaviour, he falls into disequilibrium. It is necessary to establish norms in order for him to survive. From then on, the ritual act becomes an obligation – a norm that allows purification and gradual liberation from rebirth. One has to sacrifice himself/herself in order to achieve Nirvana. What is emphasised at the level of antithetical memory is the notion of Karma – of beginning again. The end of civilisation provoked by an event of human origin gives rise to new hope. Will man recommence what he has done? The idea of permanent renewal should promote the possibility of perfection, but error is always present and possible. Man should not be trusted. It is only one step from here to speaking about rebirth. This results in the notions of Karma and Nirvana – the supreme recompense for the person who has followed the rules. Thus, the objective is to avoid man repeating the same actions.

---

27. Bhisma is the paternal great-uncle of the Pandava and their enemy cousins in the Mahabharata. He renounced marriage and the throne to allow the king, his father, to marry the woman he loved and to guarantee her the royalty of her children.

A last point, and not the least intriguing, is the current existence of an Indian polytheism based on monotheism. How do we explain such a paradox? Were the theoreticians of the Indian social system monotheists? If, as we have previously indicated, they were capable of integrating a body of scientific data of all types, including the knowledge of physiological interactions, we could imagine that they experienced the global nature of monotheism. Their unifying vision, linking all functions together by the Dharma, could correspond to a monotheistic approach. It is possible to imagine that the compartmentalisation of knowledge, due to forgetting over the centuries, was able to make the society evolve from pre-existing monotheism to a form of polytheism coloured by monotheism. Despite the important effort of memory that the Indians made, a form of forgetting appeared. The appearance of polytheism could have been born of the necessity to manage each function as well as possible, by considering man as a unit of divine origin. The gods only come to those who serve unity, which is to say of Brahman. Indians have the conviction of having gone beyond all human thought, with a feeling of timelessness and a universality of thinking that gives them openness to all the religions that serve God. Monotheism gives a homogenous and very anthrocentric vision of man. It translates a search for unity through acquired knowledge. It is curious to notice that current science, which is developed through Cartesian logic, permits us to link physics, chemistry and mathematics, and gives life a logical image. That which it was unable to encompass was confined to the domain of religion, philosophy, and so-called social sciences. Thus monotheism and a uniting principle go together without being disturbed by the irrational.

In explaining the links that exist between all things, man now only wants to have one timekeeper. Brahman, being the origin of all creation, would be kept as the pillar of antithetical memory, completed by a string of avatars working to maintain the Dharma. Thus, it is possible to integrate Hindu polytheism and monotheism.[28]

Finally, the term, 'Asura' characterises the gods of pre-Vedic India. What is really astonishing is that the status of the Asuras changed with the course of time. Originally, the Asuras were considered as just, good and charitable as they represented the Dharma. Then, a change occurred: they became proud, vain and quarrelsome when they claimed to usurp the place of the gods of the sky or to incarnate themselves as princes on earth, in order to oppress the Brahmans. From then on they represented the Adharma, and an avatara[29] of Vishnu had to come to annihilate them. Would this behaviour not provoke a war, a catastrophic event for humanity?

Let us summarise all the points that we have retained as representing antithetical memory in order to retrace the history of the civilisation which has preceded us. A very inegalitarian society was the source of social disorder. This inequality came from the existing inadequacy between the management of the economy (production), finance, ideologies, and arms, and the ecological balance in relation to eliminations. This situation created some destructive relationships. In studying the fundamental reasons for this self-destruction, it would appear that man could not have respected the conditions necessary for the physiological equilibrium of the human organism, which is to say, homeostasis. The fundamental cause would have been linked to the

---

28. Alain Danielou, *Mythes et Dieux de l'Inde* (Myths and Gods of India), Flammarion, 1994.
29. Incarnation of the supreme god, especially Vishnu, into the world in order to re-establish the weakened Dharma.

proliferation of functional prostheses, very unequally distributed throughout the society, generating great social inequalities (hypertrophies and hypotrophies). It would be a matter of a society at high risk (quadripartite) that would have developed a very large production sector responsible for an ecological disequilibrium (eliminations). Its evolution would have generated nastiness and aggression, following the manner of the Asuras. A war would have been started by man, creating a situation of extreme distress for the survivors who would have had to suffer, moreover, genetic deterioration and a phenotypical regression related, for example, with the radiation attributed to nuclear conflict

In order to accept this thesis, it is necessary to start with the theory that the survivors had a level of physiological knowledge sufficient to take into account human homeostasis and its mechanisms of adaptation. Their objectives were to have it respected. Those who survived analysed the causes of the imbalance and realised that the hypertrophy or hypotrophy of functional prostheses resulted in social inequality and was responsible for the event. After having analysed the causes, they searched for the means to prevent human society making the same mistakes. The survivors, amongst them certain theoreticians, established a normative social system so as to avoid functional excesses. To regulate the acts of each social functional representation, they created a legal system (Dharma), destined to manage the interactions between social groups. This is what we call today the 'socio-cosmic order' which could be more clearly called the environmental or eco-social interactive legal system.

Thus, all the seeds of Hinduism have been present since the establishment of the Varnas.

Theory of Survival

Everyone agrees that contemporary technical progress has considerably increased our potential power of destruction. One cannot be astonished by the reaction of Robert Oppenheimer, an exceptional man, who possessed a vast literary and scientific knowledge. At the time of the explosion of the first nuclear bomb at Los Alamos in 1945 Hubert Reeves wrote: "One imagines the exultation at the first moment of the explosion."[30]

Oppenheimer recounts how, at that instant, he recalled the words of Lord Krishna in the Mahabharata. Lines with prophetic consonances:

> The radiance of a million suns
> Explodes in a single burst in the sky
> Thus would be the splendour of the All-Powerful
> I became death,
> The destroyer of the universe.

Today, the risks of losing control are more and more imaginable, without any political system likely to change the course, or take care of itself, blinded as the states are by their own interest and profits.

Imagine that, even as you read this, you learn that man, in his madness, has provoked a cataclysm. Following a nuclear conflict, a tidal wave would, in the coming hours, submerge all the land above water. You only have a few hours to save yourself. You do not know by what stratagem and it does not matter. It is not the purpose of the argument, because you manage to survive. After some weeks, stripped

30. Hubert Reeves, *L'heure de s'enivrer, l'univers a-t-il un sens?* (The Hour to get Drunk, Does the Universe Make Sense?) Editions du Seuil, new edition 1992, p. 34.

of everything, you find yourself in an unknown land. How are you going to react?

One can thus discern three phases:

- The first could be entitled the physiological phase, where the survival of the individual is at stake.
- The second corresponds to a phase of reflection, where the causes of the event are analysed. It is at this stage that previously acquired knowledge could be identified.
- Finally, a last metaphysical phase draws lessons from the event in order to give a prospective vision of the future.

Three states that cannot be expressed at the same time but are interdependent.

In the first place, the questions you could ask yourself could be worded thus: How does one survive? What will we become? To clothe oneself, shelter oneself, drink, eat and sleep, stem from so many basic functional biological necessities. It would certainly be your vital preoccupation.

Then, having survived, you would think again about the event which, in a few moments, turned life on earth upside down, and you would seek to find out the reasons for it. Was it a natural event where man bears part of the responsibility for the disaster? If you reach the conclusion that human behaviour provoked this disaster, it would seem normal that you articulate and analyse the causes. This is the explanatory, analytical phase. How did we get there? What are the reasons that lead man to self-destruction? The notion of 'human fault' would thus give rise to conclusions about this reflection and its implications on human behaviour, which would be essential for the future of a the new humanity.

Finally, depending on the revelation that you have about the event, would you not think about the future in its entirety and try to find the solution that prevents the reproduction of an event of the same type?

What are the consequences of this event for future generations?

How do we transmit a warning message, a revelation, to prevent a similar event happening again?

How do we transmit information which questions human behaviour?

What social system should be put in place in order to regulate human behaviour, while allowing it to develop its individual and collective capacities?

Thus, by reading Indian documents, it is possible to imagine that the survivors of this disaster reacted to the three phases, which later entered into the collective memory.

As far as the first phase is concerned, eating, drinking, warming oneself, are primary objectives that are expressed in the Vedic texts, where each god fulfils a specific need. In the case of a cataclysm, vital imperatives are brought out. The reality of survival is expressed by strict renunciation and suffering.

The second phase is much more analytical. It is centred on reflection and interpretation of events. It releases feelings of human guilt, generating the vision of human sacrifice that is imposed on Indian society. If, as other texts seem to show, the event only represents the outcome of a conflict where human malice had reached its climax, the duty of memory would not be born from this distant event. It is impossible for me not to associate the image of human malice to this hypothesis, without referring to the last world war. When one has survived a horror provoked by man, how do we transmit a message, and what message do we transmit? I asked such questions to Isaac Celnikier, a Polish painter, deported to the camps of Birkenau and Buna-Flossenburg.

"Survivors, we were, and we remain witnesses to this murder, in the face of the sadism and cowardice of an era, and its hordes of hyenas denying the reality of our ashes."

Such are extracts from a text written for an exhibition of his work on the Shoah, which expressed complete distress by him every night in an outburst of anguish. "All that remains untransmittable", he said to me and it is true, he painted it!

Is it necessary to establish a difference between the human behaviour of yesterday and thousands of years ago?

There is no particular epoch for experiencing horror. Whatever it is, one invariably finds the same components of aggression: jealousy, ambition, pride, etc., as much as feelings that increase the thirst for power. Whether it is a matter of Cain killing Abel, or the Pandavas declaring war on the Kauravas, one is witnessing a fratricidal conflict based on the desire for absolute power, and the disorder that follows.

It is possible that through the real experience of a pre-Vedic period that the foundations of Vedism, which would not flourish until much later, were born. It sought proof of the gravity of the conflict. It could come from the history of men who had survived from, all the consequences provoked by the conflict, be it the effects on the environment, or the incidence of the eventual radioactive pollution on the human organism. Given the alteration of the environment, it is possible to imagine, following the example of certain physiological regulatory mechanisms, that the organism, seeking to protect part of its genetic inheritance, prevented the expression of certain genes in order to protect part of the gene pool. What consequences could be expected? The appearance of a morphological deterioration in the descendants, due to a phenotypical regression, human genetics not being able to withstand the environmental aggression, not possessing anything but the capacity to protect the genes from their non-expression.

If this hypothesis proved to be true, it would certainly be an explanation of the profound feeling of guilt still present with Indians today. Originally, only an oral message could be transmitted from century to century about the conditions for survival of an event for which man would have been in large part responsible. Through this behaviour one understood the gravity of the message.

Man would have been able to regress on a phenotypical level then, according to a better balance with the environment, start to progress again and recuperate lost functioning. It would be possible to imagine that homo sapiens and homo neanderthalis possess a common origin. On a hypothetical level, following an event of a nuclear nature, homo sapiens would have been better able to protect themselves from the effects of radiation than homo neanderthalis who would have been less able to endure them at full strength, generating mutations such as those that were responsible for their disappearance thirty thousand years ago. According to this hypothesis, Darwinian evolution would follow in its natural selection variations dependant on man's adaptation to environmental conditions, resulting, on a phenotypical level, in a sinusoid curve. The linear evolution would only retrace a limited part of the history of humanity and could only be due to a genetic re-expression of genes not expressed during this period. The ravages provoked by a nuclear war are difficult to imagine and envisage. Those provoked by the accidental explosion of the reactor at the factory in Chernobyl have been known and analysed for a long time, demonstrating the genetic changes of an epigenetic type that are able to be transmitted in the long term. Are we not threatened every day by a catastrophic conflict and are we not on the verge of a spectacular regression of civilisation and its human functions?

A third phase is prospective. It is that of hope! It presents a vision of the future and seeks methods in order to struggle effectively against high-risk behaviours. This phase will seek faithful and effective ways to send a message about an event, by explaining the causes in order to modify behaviour. And, in this context, it is above all oral transmission that can represent the best mode of information. Also, we need not be too astonished by the memorisation techniques practised in India which are still admired by our contemporaries. This fact proves the state of emergency in which the initiators of these oral messages found themselves. The objective is to work out an approach, relying on an ethic that is timeless, immutable and representative of equilibrium. Human physiology brings us a universal and timeless response from the indispensable functional balance of the body.

To imagine gods in order to satisfy elementary needs brings a response to a dramatic situation. It is by analysing the particularity of the gods that one can become aware of the events that the survivors had to manage. Supposing a nuclear disaster took place, it seems pertinent to consider the destructive side and the reparative side of the fire. Rudra emphasises the destructive side of the 'fire', and Agni, its constructive side.

Vedism would have been born in this context. The Vedic texts immortalise, by its wise and generous vision, the great human needs indispensable to vital equilibrium.

Some people would say that this vision is a work of the imagination, that we have never found significant proof of the existence of a civilisation, such as ours, being ruined.

Let us not be blind: the oral tradition is still present in India, in China and in the Himalayas. It inscribes the reality in the imagination. Thirty years ago, to suggest we would be capable of giving birth to a centaur would have been synonymous of madness, dementia. Today, madness is already in close contact with truth, and we should take another look at Egypt and Mexico, which have a cult of 'Man-Animal'!

India and China transmitted an approach to prevention by imposing it on the population. The objective of these two countries was identical: to find a way applicable to man to make him respect the fundamental balance indispensable for survival. The recipes differ: one is physiological, based on the management of homeostasis; the other relies on physical data, explaining through the representation of Yin and Yang, the existing balance between energy and matter. These two approaches to prevention represent two messages: the physiological equilibrium of the human organism must not be modified, and the harmonious balance of the energy-matter dynamic must not be perturbed.

There are truths that one does not want to hear because they pierce human pride in its self-satisfaction! Finally, supposing that a catastrophic event took place, and that man, in order to protect humanity, conceived the two systems of prevention, would that really be so annoying? Will we not be saved by imagination?

The Future in Question

It is in terms of 'behavioural regulation' that the future of our society is at stake. These terms should be immediately considered by the vision that we should have of our physiological balance as well as that of the generations to come. At the end of risk, there is life and death. How is it possible to control the search when one is incapable, through one's professional training, of expressing the slightest opinion? As Ulrich Beck explains in *Society of Risk:* "How can we, in the future, stand in the way of a potential disaster in human genetics without stifling the freedom to research, without which we are not able to live either? My response is the following: by elaborating and putting in place legal guarantees for certain possibilities of the sub-political."[31]

It is not a matter of imposing a dictatorship of regulation but of limiting, through ethical physiology, risky behaviours. Therefore, risk management is linked to putting into practice the knowledge of those belonging to a social organisation where it is impossible to separate the individual. It seems pertinent that a functional specialisation opens the way for internal criticisms, whose regulatory effects can be real. When Ulrich Beck writes, "Modern society does not have a centre of regulation,"[32] he is pointing a finger at the problem of our society. To speak about regulation is to envisage all the intervening factors in a loop, where each stakeholder has a part of responsibility for the final objective. It is necessary to specify what one means by 'regulation'.

We have seen that the establishment of the EOP was not limited to any one system of regulation. The only warning principle that can be imposed is one that in no way denies

31. Ulrich Beck, *La Société du risque, sur la voie d'une autre modernité* (The Risk Society: Towards a New Modernity) Flammarion, 1986, p. 490.

32. Ibid., p. 485.

the risks taken. It is not a matter of a principle of prevention. Each researcher in his own way develops, often enthusiastically, his idea without initially establishing ethical barriers, and not always being aware of the repercussions of his own discoveries. And ethics committees? They were created to give opinions, but their role is not to supervise researchers. They do not possess an executive power. As for legislation, it defines the action yet without defining the intended objectives. Indeed, the person responsible for the work done is the one who carries out the experiments. He/she should be capable of censuring or refusing a project. Due to specialisation, it is impossible for any other person to give an opinion in his/her place.

Favouring self-criticism in all its forms does not represent a danger, but most likely the only way of identifying the mistake before it happens.

Regulation becomes an individual problem within a collective framework, necessitating an indispensible impartiality for its authenticity. It would always be necessary to ask oneself the questions: What are the consequences of my acts? Where are the risks? It is a complex subject that requires a very good knowledge of the problems.

Physiological risk is as much at stake on the physical level as the psychic level, the one, incidentally, being totally interactive with the other. It persists on the genetic level, without our being able to intervene at all! So we should then submit!

As we have emphasised, the development of the EOP has considerably increased our functional capacities, and with them our power to act and the risks which follow.

Certain politicians are not mistaken about this in using to their advantage radio, television, and all such systems allowing them to influence the population and eventually subjugate it to their power. Business, which serves the population and manages its needs, henceforth, becomes part of a systematic component whose workings and objectives should be analysed according to the risks that they engender.

Physiology, Regulation and Medicine

The Case of French Society

The Application of Physiological Prevention to a Health System Model

It may seem odd that I include in this book a draft health system based on prevention. It is the culmination of a reflection on physiological balance and the management of various states of health. Indeed, this study owes much to the precursors of modern physiology, that of Claude Bernard and Louis Pasteur in the 19th century, W. Cannon in the 20th century and all present-day researchers who follow the same reasoning and study the regulatory mechanisms of the body. I think particularly of Pierre Corvol, Professor of Experimental Medicine at Collège de France, currently the Administrator of Collège de France, who both studies the mechanisms regulating blood pressure and angiogenesis, and Claude Hagege, Professor of Linguistic Theory at Collège de France, who helped me clarify the meaning of the term *"prevention"*.

In presenting this project, I keep the same metaphor to represent the human body as that used previously. As mentioned in the first chapter, I chose as a reference model a sphere whose representation is a circle which makes it possible to integrate all the physiological functions, to show their interactivity and rapidly incorporate a series of information. By its regular character, uniform, divisible into quarters, in a well-defined environment, it also provides support for reflection to explain the specific role of each functional component, namely, metabolic, neuropsychic, immunological and elimination, as much regarding their needs as their roles. This systemic approach makes it possible to outline in a very logical way the management of the whole, and facilitates not only medical assuming care but also the education of behavioural advice (education) and therapy. It makes it possible to combine dynamism and stability.

When the Social Security Schema was created in 1946, the preventive system was mainly destined to take charge medically of clinical risk, without the use of therapeutics. It was different from the therapeutic curative system which integrated treatment of acute and chronic illnesses and disabilities. By defining an opposition health sickness, the health system creates a bipolarity as caricatural as it is false. We shall analyse the reasons. Moreover, the dual character that it reveals gives prevention an ideological face offering health a clear and pure vision in contrast to the darkness of disease dealt with on the therapeutic level. This static conception of prevention, in its broadest sense, does not take into account on a therapeutic level functional risk factors likely to modify health and cause disease and thus provides little guarantee of health. Moreover, it does not characterise the physiological dynamics of body regulation, guarantor of healing and health.

One would have thought that the health system would evolve thanks to the advancement of knowledge acquired during the 19th century. In addition, medical laboratory techniques developed during the 20th century make it possible to specify standards in biological terms, highlighting in quantified terms the values likely to cause disease in apparently healthy individuals: in biology, it is not the cowl that makes the monk but in physiology it can still do so! The biological results make it possible today to treat functionally sound subjects, apparently healthy, whose capacities for work are unchanged, and to initiate treatments to reduce the risk of the occurrence of an acute or chronic disease. The preservation of health has become, therefore, dependent on taking medication without the disease being declared. However, this new field of medicine not being individualised on an economical plan and financed in the health system, the legislature has decided that therapies for primary prevention be integrated into the therapeutic management of secondary prevention (acute diseases) and tertiary sector (chronic diseases) by the Social Security, which is conceptually self-contradictory because the person is not considered sick. He/she treats himself/herself, however, so as not be ill!

Take an anti-cholesterol treatment, such as a statin: does it mean that a person is suffering from a disease or has he/she intended only to protect himself/herself against the occurrence of a cardiovascular accident like myocardial infarction? If we consider that curative treatment is considered as part of our regulatory mechanisms, it is an act of primary prevention.

Assuming that our body is a system of prevention, we understand therefore that the curative treatment, whose content varies depending on the state of health, is integrated into all levels of prevention regardless of health status. There remains three different levels of therapeutic management, depending on the condition of health. It should be necessary to adapt treatment to each of these three levels of prevention.

This situation led me, in 1999, to specify the domain of the definition of prevention. I forwarded to the Academy of Medicine a draft communication on the concept of prevention co-authored with Professor Claude Hagège. This communication is essential to the comprehension of the project: "It may seem surprising that a doctor requires the assistance of a linguist to explain his proposals in an area that is apparently far removed from linguistics. This is because the field in question uses a certain terminology, whose implications are serious enough to require an evaluation of the meaning of the words."

The term that is central to the debate is 'prevention'. In its current usage it seems that the meaning attributed to it does not include the notion of regulation, in the sense of the prolonged and control establishment of balance between biological factors. Yet it was the central inspiration of Claude Bernard when he emphasised in 1865, in his *Introduction to the Study of Experimental Medicine*, the importance of these mechanisms, which maintain "the unity of the conditions of life", in other words, the importance of what Walter Cannon, in 1929, was to call 'homeostasis'. The improvement of the vaccine by Pasteur in 1865 meant that he too was part of the long series of efforts which biological research is accomplishing to heighten the forces of balance in the human body.

However, because of traditional teaching, the general practice of medicine today is inspired, explicitly or not, on a dualism which separates the preventive from the curative. Given this proposition, it is important to remember that prevention is initially the same as what our body does naturally, and that we call it 'natural prevention'. It is because this basic data provides clues about the state of balance of the body that another form of prevention exists, namely, 'voluntary prevention'.

This is precisely where medicine intervenes. But voluntary prevention, by the very fact that it is about the intervention in the service of what is natural, necessarily implies behaviour, and consequently, it is situated in an essentially social and ethical framework, that of the well-being of human societies.

That is why we can no longer limit ourselves to acting after the phenomena whereas it would be advisable to act before position. If we perceive the importance of this change of perspective, then overlooking the epidemiological definition of the three levels of prevention (primary, secondary and tertiary), such as that established by the World Health Organisation (WHO), we will define, this time in terms of clearly physiological, three states of health:

- a primary state, that of apparent "good health", which is, in fact, only the state of homeostatic regulation resulting from the permanent struggle of our immune defence system against the aggressions of the environment and hazards of internal evolution;
- a secondary state of health, that of acute illness; and
- a tertiary state of health, that of chronic illness.

The use of cybernetic systems here can help develop a topological model to properly classify the physiopathological facts. It results logically from what precedes that the curative aspect of prevention is a basic fact, the importance of which was not sufficiently perceived up to now. If indeed we want to think freely, then we realise that the curative act is, in reality, only conceivable when integrated in the preventive act.

Today, the study of disease has changed. It is in the hidden domain of primary prevention that basic and epidemiological research anticipates the pathological clinical signs by studying the risk factors, moving upstream the therapeutic objectives.

"The abyss underlying evil and which was evil itself has just emerged in the light of language."[1]

Henceforth, it would have to be considered, against established usage, that we cannot speak about health without talking about prevention, and therefore the Ministry of Health should be called the Ministry of Prevention. If we consider that prevention is nothing other than health itself, then it may appear that there is through the content analysis of words, a radically new view on society which is proposed here.

Undoubtedly, it is important to give a certain audience to reflect on the appropriate relation between health and prevention. Obviously, it is about a debate which is not only medical but also cultural and political. At stake is the human being, and his/her integrity.

The WHO had the merit to use the word 'prevention' in a global sense to indicate assuming responsibility for all states of health, whether it is about the incidence of a disease (primary prevention), about the prevalence of a disease (secondary prevention) or about chronic diseases and

---

1. Michel Foucault, *Naissance de la clinique* (Birth of the Clinic), 1st edition: 1963. Reprint of 7th edition "quadrige", February 2007.

disabilities (tertiary prevention). This epidemiological definition of prevention integrates all forms of prevention, that is the one determined by the body itself and the one which we do have voluntarily to protect ourselves. For example, vaccination uses the body itself, by its own defences (primary prevention), while antibiotic therapy combines with the immune system to fight against a microbial infection (secondary prevention). In the first case, it is the body alone which deals with the prevention, and we know the current expansion of vaccines which are even used in oncology. In the other, it is the drug which adds its capacities to those of the body.

Unfortunately, the word 'prevention' is often used to indicate different objectives. To add to the confusion, the classification of prevention in public health, although at three levels, in fact, creates a dualism between the non-curative act (primary prevention) and the curative act (secondary and tertiary preventions); primary prevention ultimately being identified with overall prevention. The expression "prevention is better than cure" contributes to this bipolar view, underlying the idea that to prevent without treatment can avoid the fact of treating, and of curing.

Thus, by its conventional structure, the classification of prevention in public health does not cover the same territory as that used in physiology. Physiological prevention, which takes into account the regulatory mechanisms of the body, does not establish any gap between primary and secondary prevention.

Take the example of the book titled *Social and Human Science, Man, Society and Medicine* where G. Lazorthes defines preventive medicine as the medicine of the healthy man. Then, prolonging his reflection in a chapter named "The Prevention of Diseases", he tells us: "Instituted throughout life, prevention stops many diseases which appear with age" and cites in disease prevention: "A well-balanced diet, physical activity, oral hygiene, the fight against smoking, alcoholism and industrial pollution, limiting speed on the roads, routine screening, epidemiology and health education."

What is meant by a healthy man?

Unquestionably, nutritional behaviour, physical activity and hygiene are factors of physiological balance. However, speaking of "routine screening" demands a precise definition. If it is about biological risk factors, we are situated in primary prevention. If this screening reveals a non-detected disease, then we are in the field of secondary prevention. In any case, the author of the book does not seem to consider that the therapeutic management of risk factors is an act of primary prevention, the purpose of which is, however, "to eliminate many diseases"!

The following was noted by Jacques Chirac, President of France, in a letter dated 24th October, 2002, following my presentation of this project to Madam Marie-Claire Carrère-Gée, Social Advisor to the President of France:

> In establishing the curative dimension of prevention, you manage to replace the prevention from the core of the health policy and the healthcare system.
>
> Madam Marie-Claire Carrère-Gée, who follows these matters in my office told me that she has met with you recently. She emphasised the coherence of your project, on which you have been working for about ten years ...

The physiological view of prevention effectively presents the quality "to establish the curative dimension of prevention".

Indeed, this simplistic distinction between "non-prescribing doctor of prevention" and "prescribing curative doctor" is becoming increasingly unsuitable in medical practice. Today, the fight against risk factors is essential. However, implementation has not yet been found.

If our politicians were more aware of what is at stake with "primary prevention medicine", in the physiological sense of the term, they would specifically deal with primary prevention therapies as well as some recently delisted medicines, thus avoiding casting discredit on the medical profession, bad prescribers because they prescribe medicines "known as unnecessary". Imagine the thousands of doctors, having received marketing authorisation (AMM), suddenly being accused of prescribing unnecessary medicines! Imagine the patient's surprise when he/she learns that his/her doctor has prescribed only inefficient medicines, and thus useless! Imagine the anger of researchers who have worked with conscientiousness on molecules which are affirmed today as being useless! The domain of the definition of primary prevention remains completely misunderstood today. All these decisions could have been avoided if public authorities had accepted the idea that there exists today, in reality, a tripartition of medicine and that certain medicines, perhaps less obviously active than others, but effective nonetheless, should be recognised as medicines of primary prevention. If we followed this reasoning, thousands of jobs could be saved, because the recognition of primary prevention and its treatment by a marketing authority in primary prevention (MAPP) giving the primary prevention medical act its proper importance, even if reimbursements of medicines are very low. If the development of primary prevention cannot be included in the control of acute and chronic diseases, it must be recognised and monitored by health authorities.

A health system which addresses an entire population must be simple, logical, coherent and comprehensible to each and every one. According to what we stated in the first chapter of the book, the human body is a system of prevention which we control, whatever the state of health. That is why, we must distinguish between the state of good health, the state of acute disease and the state of chronic disease and disability. The Ministry of Prevention should associate itself with the Ministry of Preventive Medicine and the Ministry of the Environment, as I have stated in *La lettre du Collège de France*[2].

As I have already mentioned, it was following a letter I sent to President Jacques Chirac on the 15th July 2002, which I had the privilege to present to Social Advisor to the President of France, the project of a health system based on prevention:

> During the televised speech of July 14, 2002, you stressed the importance of prevention both individual (cancer) and environmental (Johannesburg conference). For several years I have maintained that France has a real need to change its health system by basing it on prevention. In 1993, I developed a project which received the attention of the general director of health and which I presented to Mr Lavergne, Social Advisor to the President of France, François Mitterrand.

---

2. *La lettre du Collège de France* (The letter of Collège de France) n° 2, May 2001.

This project aims to change the behaviour of the French regarding disease. Its foundations are simple. There are two expressions of prevention in humans: one autonomous or innate related to our physiological constitution (immune defence system, etc.), the other voluntary, reflecting the activity necessary to manage independent prevention.

While being based on physiology which expresses our capacity for action, it is possible to define three levels of prevention exactly in keeping with the epidemiological classification of WHO. Thus, we end up with a tripartite division of medical practice instead of the bipartition currently in force (dualism between the preventive system and the curative system).

Its application would provide a durable solution to the balancing of the budget of social security while being based on an analytical method which I named VISA (Voluntary, Incentive, Solidarity, and Anticipation) logically combining the individual action and that of management, while preserving state's role of permanent arbitrator.

Thus I stress, Mr President, the need for defining prevention within a broader context including the management of different health conditions in relation to the environment. These ideas have led to the presentation of a draft communication on prevention presented to the permanent secretary of the Academy of Medicine thanks to the support of Professors Claude Hagège and Pierre Corvol, member of the Academy of Sciences. They have also been diffused in the pages of Quotidien du Médecin last year ...

Diagram 4

The Project of a Health System Based on Prevention

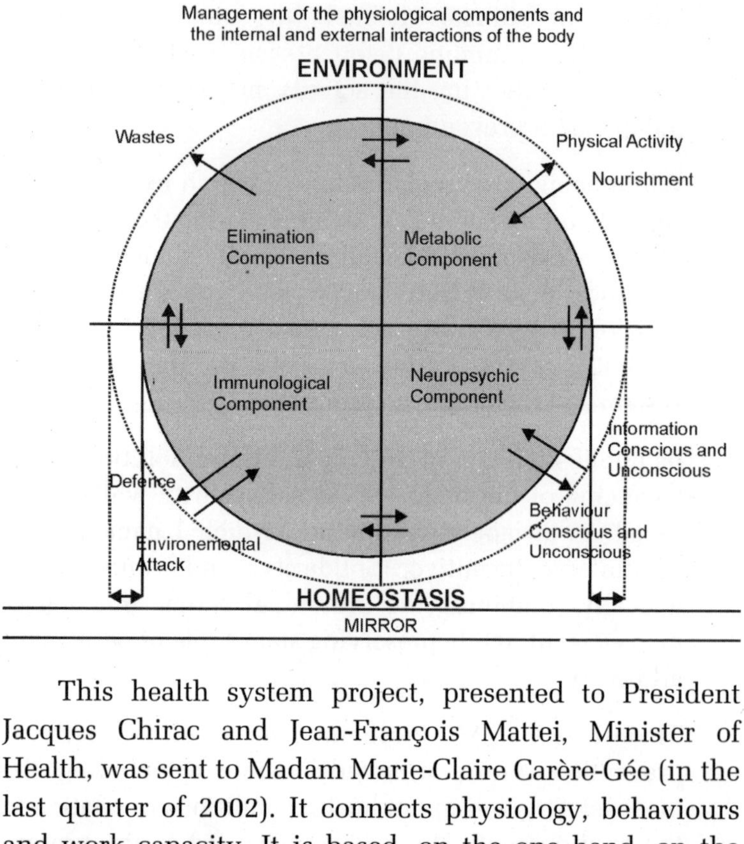

This health system project, presented to President Jacques Chirac and Jean-François Mattei, Minister of Health, was sent to Madam Marie-Claire Carère-Gée (in the last quarter of 2002). It connects physiology, behaviours and work capacity. It is based, on the one hand, on the recognition of the functional role of the physiology of the human body, itself dependent on regulatory mechanisms which constitute innate prevention, and on the other hand, on voluntary participation which has an objective to respect the physiological balance according to needs and the environment.

Diagram 5

Modelled Representation of Variations of the State of Health

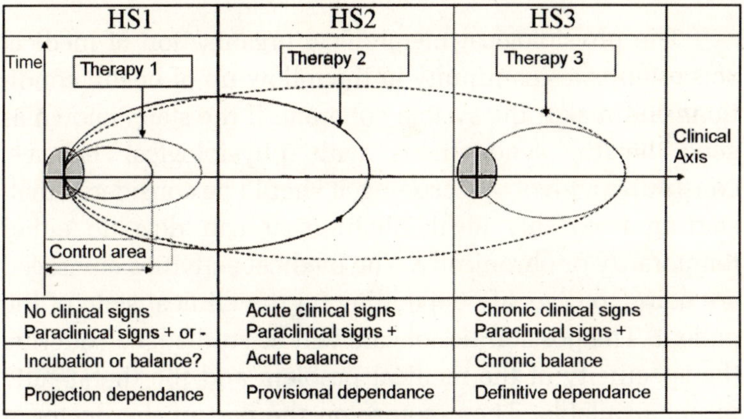

Dr Bernard JOUANJEAN
Health System Project

By linking behavioural physiology to behaviour and behaviour to budgetisation, this project gives a sense of responsibility to the people and the doctors.

"Physiology (gr. Physis-nature; logos-speech) is part of biology whose aim is to study the functions and properties of the organs and tissues of living beings."[3]

It is a question of eventually bringing doctors to a trifunctional and three-dimensional organisation. The advantage of this system is to differentiate all types of medical interventions which are today profoundly intricate and multifaceted. We can distinguish three categories of acts, namely, those which aim to prevent the occurrence of an acute or chronic disease involving management of several risk factors not clinically expressed (non-event), those requiring immediate care due to the urgency of a situation

3. *Dictionnaire des termes de médecine* (Dictionary of Medical Terms), 22nd Edition, Garnier/Delamare, M9-aloine, 1989.

and finally those acts that are predictable in the long term and where the patient has a permanent dependency with respect to a treatment or a third person.

The physiological hierarchical organisation of medical acts establishes continuity in the follow-up of health conditions and makes the system coherent. If the state, known as good health, synonymous with physiological balance, constitutes a pivot of reasoning, it should be considered that, starting from this ideal situation, it can degrade either temporarily or chronically. The medical activities envisaged are going to diversify depending on the clinical state of the patient. The possibilities of care will be studied, considering the specificity of the medical problem and the therapeutic means available. They require on the part of the doctor, a continual and conditioned adaptation of his knowledge.

In this functional tripartition the conditions of health must correspond, point by point, to a tripartition of the medical acts which require a distribution of information and action according to the needs of the population.

Furthermore, the medical profession should, through its members, speak the same language. 'Prevention' is a commonly used word, almost generic, which is assigned variable fields of definition, as it is possible to observe in medicine: whether it is its epidemiological classification established by WHO, used in public health, a physiological classification or other possible classifications, like biological, genetic, etc. In order to satisfy a principle of coherence, it seems essential to choose a definition of prevention acceptable by all, which can be applied universally throughout the health system.

We believe that physiology, through contributions to the state of health and body functions, gives an innate vision of prevention and management. It can constitute a solid base and progress to a health system built on tripartition. Also, the physiological approach to prevention presents, in France, a historic character because, since Claude Bernard, all the physiology researchers work in particular on the mechanisms of regulation of the body. It defines the universally accepted laws of regulation.

Further, it constitutes not only an opening in fundamental research, but also provides pharmaceutical laboratories an explanatory support in the presentation of some of their products. On a practical level, it should be noted that a physiological classification is already used in medicine (cardiology). By bridging the gap between biology and clinical expression, it allows the establishment of links between health status and capacity of action and the budgeting system. It makes it possible to set behavioural objectives (research of the state of balance) and to change individual and collective behaviour in a positive way.

There is currently no official physiological classification of health conditions to determine the method for management control. In adopting a physiological classification, at three levels, we place ourselves ahead of the one proposed by the World Health Organisation (WHO) several years ago on an epidemiologic level.

Physiological prevention can be defined as the management of the regulatory mechanisms of the body, permitting to maintain, cure and improve a clinical situation in order to obtain a stable state of health as close as possible to its balanced state.

Physiology makes it possible to distinguish three particular medical situations, facilitating a modelling of the health system and, consequently, its budgeting. The primary state of health (SH1) defines a health state without functional impairment. The secondary state of health (SH2) characterises the deterioration of physiological functions either because an internal cause has changed the physiological situation gradually, and has then developed suddenly (myocardial infarction), or because of an external aggression, physical, biological, chemical (burns). In all these cases, the body develops a system of defence and repair to which is sometimes grafted a therapeutic aid. After a period of clinical expression, the body returns to a new state of balance, either spontaneously or through therapeutic aid. This is a temporary deterioration of health. So, angina, acute bronchitis, a fracture or an uncomplicated myocardial infarction all constitute temporary conditions that can be evaluated over time. This point should be noted because it is essential for evaluating the budget at this level.

When the body presents deterioration to the point where it is incapable of returning to its primary state of balance, or in other words with all its functionalities, there is a lasting imbalance indicating a chronic condition or disability. Several classes of conditions are concerned: diseases requiring replacement therapy (thyroid deficiency, diabetes type 1), chronic conditions which are treated symptomatically (high blood pressure), physical attacks with persistent disability, etc. All these situations come under the name of tertiary state of health (SH3).

The adoption of a tripartition health system allows, during the study of the evolution of diseases, the conservation of medical monitoring and care which is logical and without hiatus. In the beginning of the 21st century, several diseases are announced as major public health issues and provide an example of an area where the system can be implemented: diabetes, cancer, cardiovascular diseases and AIDS[4]. Each of these diseases has the characteristic of

4. AIDS (Acquired Immune Deficiency Syndrome)

evolving according to a group of risk factors preceding their occurrence. These diseases called 'multifactorial' develop over time according to genetics, lifestyle, environment, etc. It is possible to comprehend the disease by following its development and seeking to limit its expression in all three stages we have defined.

Therefore, we will choose the treatment best suited to the conditions present. The establishment of a tripartite model of the health system brings clarification of medical treatment and care of patients, makes the responsibility of doctors more coherent, will change the behaviour of patients, avoiding excesses of information on risks (litigation, etc.) and, consequently, to clarify its budgeting.

To better understand the links existing between the three states of health, it is useful to see the diagram given in the text. In schema 4, we have represented three domains corresponding to the states of health SH1, SH2, SH3, and have made separation zones between these different levels. The separation between level 1 and 2 is complex. If a rich clinical semiology perfectly fits level 2, a set of isolated paraclinical results can raise the question of which stage it is considered to be at (elevated uric acid, cholesterol). In fact, there is no mandatory subordination between risk factors and pathology. We must consider their specificity and the significance which must be assigned to them. Depending on the potential seriousness, it is possible to define according to their intensity of risk R1, R2, R3, which requires different clinical and paraclinical monitoring. For example, screening for hypercholesterolemia may represent a risk R1 while the discovery of a viral antigen or specific antibody (R3) can have an unfavourable significance in the short term (seropositivity for hepatitis C, the AIDS virus, etc.). Thus, the risks are concurrently related to the time lag between paraclinical diagnosis and the clinical expression of a disease and with its severity, which we must know how to anticipate. This field of intervention is unpredictable. This random area, functionally silent, must be specifically supported from paraclinical screening onwards. It is the epidemiology which will determine the incubation time, predictable based on records of clinical observations and biological investigations.

Medical monitoring in the field named 'unpredictable' must be carried out by a hospital internist (secondary prevention doctor) and by a doctor of primary prevention, who co-prescribe biological tests and possible treatments. For example, an HIV positive subject (of AIDS) must be followed by a hospital clinician and a doctor of primary prevention who will deal with the clinical and paraclinical monitoring of the subject. This period of pre-clinical uncertainty is accompanied by monitoring the duration, which is based on the occurrence of physiological changes. The doctor of primary prevention is thus the assistant of the hospital clinician and must be attached to the hospitals.

The separation between levels 2 and 3 is easier to determine. Indeed, an acute disease has, by definition, a clearly defined duration. It can be easily assessed, thanks to epidemiological studies. The disease becomes chronic as soon as it lasts more than the statistical estimate. The cure of patients at level 2 (provisional dependence) is limited in time. On the basis of the principle that the final dependence (level 3) is specifically dealt with, we will see that the budgeting of level 2 is made easier.

The separation zones determine the budgeting of the system.

According to the preceding reasoning, medical practice has three different and complementary professions: primary prevention medicine or preventive health medicine, secondary preventive medicine or medical prevention of acute disease and emergency, tertiary preventive medicine or preventive medicine of chronic diseases and disabilities.

The practice of primary preventive medicine displays the most innovative features of the system. To properly understand its role, it is necessary to refer to the two diagrams. It is initially a representation of the unstable equilibrium of the body (Diagram 4). The latter, by virtue of its regulatory mechanisms, tends to continuously return to a state of balance (homeostasis). In order for this regulation to operate normally, it is necessary that information of internal and external origin be managed and followed by a regulating action, thus permitting to establish a stability of our state of health (elliptic curves of Diagram Diagram 5). For example, an antigenic aggression (bacteria, virus, etc.) will be followed by an immune response, allowing the body to protect and store the antigen. Vaccinations use this principle. At this stage, nothing is apparent and the individual appears in good health while the immune defence mechanisms are activated. Until now, medical prevention played no direct healing role in these defence mechanisms. We believe that, given knowledge in immunology, a primary prevention medical doctor must be prescriber/co-prescriber of the therapies for stage 1. This care can be operational today. We will reconsider the subject later.

Primary medical prevention is responsible for assuring the balance of the physiological functions of the body. It manages all the physiological needs in order to ensure good health. It is also responsible for managing health to whom can be assigned four major roles. Firstly, he/she is an educator of general practitioners (P1a) and specialist educator (P1b). He/she is a very important player in the detection of risk factors and diseases. I consider equally that it must be a prescriber (P1a) and co-prescriber for pre-pathological states (P1b). Moreover, it plays a key role in epidemiological investigations.

"To educate" is not only to teach, but also to support the individuals concerned. Above all, medical education must establish the state of affairs by analysing the lifestyles and habits (sociological implications) of the subjects in order to give advice to change behaviours, not only of children, adolescents and parents (importance of the interaction child-parent) but also of adults who are looking to achieve a state of balance. This implies a significant investment of time listening and analysing, a good knowledge and understanding of the social environment to facilitate the integration of prevention ideas in everyday life, a good formulation of objectives to be reached and a discourse simple to understand and easy to apply.

By general practitioner education (P1a), we mean all education done within the framework of schools in a non-pathological context. The specific education (P1b) concerns the coverage of individuals at risk (obesity/diabetes, high cholesterol/vascular disease, alcoholism/cirrhosis, smoking/lung cancer, etc.).

The education of general practitioners relates to the explanation of the role of the four major physiological functions and their interactions. They are: nutrition (metabolic function), individual and social psychology (neuropsychological function) which aims to study individual and family lifestyle and the sources of imbalance, good management of immune functions against environmental aggressions (vaccinations) as well as the elimination of waste through the intestines, kidneys, lungs and skin. Managing the interactions between the components, such as neuropsychologic, metabolic, immune, elimination and environmental, is highlighted, thanks to behavioural counselling and in particular each individual should be encouraged in a physical activity.

Taking in charge the general education of children (P1a), the management of organic needs varies from infancy to adulthood. The paediatrician plays this role in early childhood until school age. Subsequently, medical monitoring, which should be regular, is often neglected. It is necessary to make parents responsible for the health of their children. We suggest the establishment of a primary prevention contract with families called "health contract" encouraging parents to keep their children supervised by a clinic or a free surgery. A primary prevention notebook or "notebook of health advice" (not to be confused with the "health record") would be presented annually to the preventive doctor. He would certify that consultations have been effected so as benefit from family allowances. This would not be a notebook for recording medical information. A specific orientation could, however, be annotated to the need of consultation to a dentist, nutritionist, psychologist, etc. Only the orientation would be indicated in the notebook, but a letter given to the subject for the attending physician could indicate the causes of this orientation.

We have seen that the primary prevention doctor is in the upstream of clinical pathological situations. We will take the example of the history of diabetes, which represents an example of specific nutritional education (P1b). The letter "Prevalence" of October 2002 of the Institute of Health Surveillance has a very evocative title: *Diabetes: evil of the century?* With two million patients registered in France, diabetes, especially type 2, is announced to be one of the major challenges of the 21st century. It is linked to poor food hygiene, particularly to a diet too high in fat for a sedentary lifestyle due to lack physical activity or to genetic factors. Type 2 diabetes is highly related to the current lifestyles of Western societies or those who adopt it gradually. It is entirely possible that stress plays a part in the emergence of the disease. Diabetes prevention involves changing habits. The multiplicity of obesity in children leads to both its expansion and greater precocity. The first national survey conducted by the CNAMTS in 1998 revealed a prevalence rate of 3 per cent with results even more alarming in the DOM-TOM. The number of undetected diabetic patients would thus be between 300,000 and 800,000. The prevalence of this disease, which will increase in coming years, is also linked to both the ageing population and changing lifestyle. This epidemiological pattern highlights the need for educational control of diabetes. We must change the behaviours of children and encourage parents and children to change some lifestyle habits. Primary prevention doctors can play a major role in the care of obese young children before the onset of diabetes and epidemiological monitoring, trying to educate them and their parents.

The primary prevention doctor divides his/her liberal and/or paid activity between school, his/her practice or clinic and the hospital to which he/she is attached. We follow the same plan as previously: a general screening and directed testing. The general screening is done as part of paraclinical systematic checks. It consists of establishing an inventory at a given moment in the life of an individual. It sometimes leads to highlighting an obvious disease, diabetes for example, or simply directing biological monitoring more specifically. Subjects at risk are detected either during systematic examinations, or because of a personal request. The clinical and paraclinical results are compared with an epidemiological data bank, which establishes the division between primary and secondary prevention. The intermediate zone between them is called the random zone, which corresponds to an epidemiologic risk level requiring a double management of the potential patient (who is not yet a patient in the clinical sense of the term) given the risks.

The primary prevention doctor can be led, within the framework of his/her daily work, to prescribe treatments adapted to a non-pathological functional state. We believe that certain non-reimbursed medicines may be prescribed within this framework, whether antioxidants, vitamins, magnesium, etc. It would be necessary to establish product quality monitoring, resulting in a marketing authorisation of therapies for primary prevention to avoid any form of abuse. We believe it is useful to create, in each hospital department (department of infectious diseases, department of diabetology, oncology department, etc.), a "consultation of epidemiological surveillance", integrating clinician and primary prevention doctor. In addition to other objectives, the purpose of this consultation is to treat pre-pathological clinical states likely to degenerate. Its specificity is to provide treatment to individuals who are functionally healthy, but with risk factors, who could benefit from therapeutic care.

The subjects potentially at risk are monitored during hospital visits by the primary prevention doctor (co-prescriber) and secondary prevention hospital doctors (acute disease doctor) who implement and prescribe first-line treatment. They establish the frequency of consultations according to clinical risk assessment, biological and iatrogenic. They update the epidemiological data. Then, the primary prevention doctor, who is present during the consultation at the hospital (as an attached doctor), deals with the prescribed treatments: he/she is a co-therapist who, during the period of treatment, becomes a co-prescriber. He sets up periodic monitoring of the patients at risk, outside the hospital. He/she informs the clinician in the event of changing results. The period between consultations is determined by the hospital clinicians. The primary prevention doctor keeps a record of the results of the biological analyses, which subsequently will be transmitted to an epidemiological data bank. Before considering the epidemiological role of a primary prevention doctor, we must specify the role of epidemiology in the health system. It is impossible to manage a tripartite health care system without founding a relatively complex epidemiological system. In particular, it is necessary to establish an epidemiological data bank. The participants of this bank are situated at several levels. The first level is comprised of the doctors of primary, secondary and tertiary prevention. At the second level are epidemiology research organisations (National Sanitary Surveillance Institute, INSERM[5], etc.). Hospitals must be the first point of information convergence, and transmit the results to an "integrated results centre". A commission of medical experts (in their field epidemiology) is responsible for establishing the separations between primary and secondary prevention, and between secondary and tertiary prevention. The epidemiologic data bank thus makes it possible to define the amounts to be budgeted at each level.

Secondary prevention medicine involves the management of acute diseases and was described and commented on previously. The role of the secondary prevention doctor was also specified. Moreover, we can say that he/she is also a field doctor (emergency physician, "disasters" doctor, etc.) and a hospital internist (clinical diagnosis and therapeutic care of undiagnosed diseases and hospital follow-up in the acute phase).

---

5. INSERM: *Institut national de la santé et de la recherche médicale* (National Health and Medical Research Institute) (France) 166.

The tertiary health condition defines situations of chronic diseases and disabilities calling upon physical and/or biological prostheses, definitive therapies marking the dependence of the individual in view of this situation (tertiary treatment). We will specifically tackle the problem of dependence in the development of this chapter.

Some diseases can become chronic and so their treatments pose real problems. Some care resources, such as the ANCRED[6] network, are good examples of an improvement in the treatment of patients with type 2 diabetes, thanks to the treatment networks of the city. These networks have been created to overcome the shortcomings of the health care system. Indeed, the social security health insurance refunds all that is acute (hospitalisation, consultations, drugs) but does not support patient education or prevention of complications, including dietary advice by doctors and paramedics. Furthermore, 90 per cent of type 2 diabetics see general practitioners, who are poorly prepared to support these patients who require a particular follow-up, and who must learn how to change their lifestyle. The diabetes network has assumed a mission of education in city medicine and committees have worked on cooperative protocols and data sheets to assist practitioners to follow the recommendations of the ANAES[7] and AFSSAPS[8]. This situation shows the confusion existing between an assumption of responsibility in secondary prevention and an assumption of responsibility in tertiary or even primary prevention.

How can this confusion of roles be avoided? In the tripartite system, the hospital would become an interactive

---

6. ANCRED: *Association nationale de coordination des réseaux diabètes* (National Association for Coordination of Diabetes Network) (France).
7. ANAES: National Agency of Health Accreditation and Evaluation.
8. AFSSAPS: French Agency for the Safety of Health Products.

centre, a place of exchange and training among doctors of levels 1, 2 and 3. The hospital would record all the clinical and paraclinical results in order to be send them to the central epidemiologic coordination service which would integrate these results. On a practical level, each hospital service would have an epidemiological follow-up consultation specific to each speciality presenting risk factors, for example, for diabetology, cardiology, infectology, oncology, etc. There would be three monitoring levels related to epidemiological risks: the management of risk factors without affecting functional impact, management of risks related to acute diseases, and management of risks related to the chronicity of a disease, and resulting in the dependence that follows whether it is about the rehabilitation of disabilities or the daily care generated by this situation.

On a theoretical level, the tripartite system has the advantage of characterising different and complementary medical exercises. We recommend this tripartition be carried out for a trial period by combining areas from primary prevention through to tertiary prevention. We would thus isolate the secondary prevention in its hospital activity or in the field, particularly in situations requiring immediate therapeutic care.

Some of the objectives of prevention are shared. Education is at the heart of the problem.

Take the example of diabetes again. In primary prevention, the individual causing concern is the one who risks becoming diabetic: it may be an obese child who has a diet too rich in fat and who lacks physical activity. If he/she does not change his/her habits, he/she will become diabetic. In tertiary prevention, the subject is an acknowledged diabetic, but in type 2 diabetes we find mostly the same risk factors. The record of health advice would indicate the patient in apparent good health and the patient affected with chronic illness or disability. It would thus be need as an objective of regulation of the state of health in both cases. The educational crossover of these two entities should be in the hospital in the "services of epidemiological monitoring".

Screening for risk factors may take place either during routine paraclinical examinations (primary prevention) or because of an altered state of health (secondary prevention). In primary prevention, the discovery of risk factors may not in itself constitute an obligation of health care. However, at this stage, it is necessary to educate in order to change behaviours. Other risk factors, independent of behaviour, may also be pathogenic. It is obvious that the discovery of an HIV-positive case cannot be compared to that of cholesterol. Clinical and paraclinical monitoring varies because the scale of risks differs considerably. As an HIV-positive subject can keep the same functional capacity as a healthy subject, he/she enters, according to the physiological classification of prevention, within the framework of care management by those involved in primary prevention. It is necessary to differentiate the risk factors according to the epidemiological seriousness of the disease. We assess risks depending on health status (R1, R2, R3), from where risk factors can be better managed. The consequences in the short or long term of a particular factor are studied according to the clinical expression, the acquired medical knowledge and the applicable treatment. Only epidemiological context provides guidance on what to do.

We have already considered the crucial role played by epidemiology in the health system, specified the various actors (doctors of the three levels of prevention, research organisations and hospitals) and shown the importance of the creation of a "central bank of epidemiologic data", this data being periodically updated according to knowledge acquired.

In this tripartite system, the hospital becomes an interactive centre, a place of exchange and training. It is at the hospital that all the clinical and paraclinical results are recorded and transmitted to the "central epidemiological coordination service", to come under the control of the central bank integrating all the results. We believe, it is useful to create, in each hospital department (department of infectious disease, department of diabetology, oncology department, etc.) an "epidemiological consultation surveillance" integrating the joint action of a hospital clinician and a primary prevention doctor. The purpose of this consultation would be to treat the clinical pre-pathological conditions likely to degenerate. Its specificity is to provide treatment for functionally healthy individuals having risk factors, who may benefit from therapeutic care.

The epidemiological bank must be central to the health system. The management of primary prevention is possible only when the separation between levels 1 and 2 is established. This is why it appears necessary that a meeting of the epidemiological bank, assembling French and international data, should take place each year to determine the conditions of care in primary and secondary prevention re-examined annually.

As mentioned earlier, the hospital is an interactive centre, a place of training and exchange between practitioners of the three levels. The creation of specific "epidemiological follow-up consultations (EFC)" would allow the management of corresponding treatments and it would act as a buffer between the three levels.

It facilitates epidemiological investigations, integrating the results of clinical research and therapy in medical practice (INSERM link – hospital service), the accountability of hospital doctors – primary prevention doctors in the medical monitoring of patients, and, consequently, a complementing of care management.

The reform thus proposed would have a profound impact on medical studies. Indeed, students would have to study within the tripartite framework. The doctor would then specialise according to the choice of practical work he/she has done. To make this choice, he/she will have to pass the corresponding certificates.

- The specialisations of primary prevention medicine would be physiology, paraclinical and clinical immunology, epidemiology and public health, screening by paraclinical monitoring and testing (ECG, spirometry, etc.), professional and non-professional environmental medicine and behavioural toxicology (alcoholism, smoking addiction, poisoning by drugs).
- The specialisations of secondary prevention medicine would be internal medicine – clinical and therapeutic, surgical pathology, secondary preventive medicine, oncology, secondary therapy, medical emergencies and disasters (compulsory SAMU[9] training) and epidemiology.
- The specialisation of tertiary prevention medicine would be gerontology, clinical epidemiology, neurology (Alzheimer's, dementia, etc.) and psychiatry, physical therapy and rehabilitation, nutrition and toxicology.

Regarding the financing of the system, our objective is not to quantify the cost of each level but to demonstrate that it is possible to budget the system according to the

9. *Service d'aide médicale urgente* (medical emergency service).

considered health status. The physiological laws form the basis of this health system, as they determine both the medical and financial management. The reasoning adopted in the financial plan is as follows: we distinguish in physiology three health states corresponding to three capacities for action and thus three very different behavioural situations. Several questions must be asked for each individual at each level of health: Can his/her state of health be anticipated? What are his/her capacities for action and work? What can he/she achieve voluntarily? Is it possible to encourage him/her to deal with it himself/herself? Is the community involved? Should it be supportive of his/her condition? Which community is it about?

A fundamental principle must be applied: we must make all those involved responsible for the management of the health system – the individual himself, the insurance companies and the community – and not increase the charges.

We begin with the principle that explaining to each individual the fact that he/she is subject to regulatory mechanisms, to which he/she owes his/her stability, makes it possible to enhance each act of prevention and to give a sense of responsibility to each actor. Whatever the health status, the intellectual approach adopted encourages greater investment in more responsible behaviour. Also, to tell someone "you were born with a system of prevention and I, the doctor, am a manager of this prevention" involves a deep awakening and a reflection on what we believe to be and what we are. When it is about an individual in good health, everything can seem simple. If it is the case of a disabled individual or one suffering from an incurable disease, the assertion is difficult. Yet, the acquired knowledge and technical advances made in patient care help us become aware of developments in medicine. We may one day treat some illnesses which today are called "incurable". Our society works on this type of injustice and tries to compensate for them with defined therapies but the organisation of tertiary prevention leaves a lot to be desired. Prevention cannot be isolated in a stranglehold of perfection. If so, it would be identified with an ideological system rejecting imperfection. However, the regulatory mechanisms of the body are always active regardless of health status, even if one is obliged to compensate for a lack or a drift by a mechanical or biological prosthesis.

The physiological laws governing the body are identical for all subjects, whether disabled or healthy. However, in the event of dependence, certain situations appear to be deeply unjust and, at the same time, difficult to support for the people involved. This support is a heavy responsibility both financially and in human terms. Moreover, it puts a strain on the freedom of each person. We must learn to manage dependence. To do this, it is necessary to anticipate this possibility both on community and financial levels. Family policy must be central to this problem and in this context the tertiary prevention doctor is a pillar of this support. Today, everyone is at the mercy of dependence, especially due to ageing population. We think that families, groups of friends and associations must invest in the creation of "family dependency homes". This would avoid the isolation of people with disabilities and facilitate their care.

Some rule must be applied: greater clarity and insight into how to manage the body, taking into account the enhancement of its capacities to adjust and regulate itself, more individual medical responsibility at each stage, better management of risk factors, a valorisation of individual and collective voluntary participation, systems of individual and collective incentives, encouraging the capacity to anticipate health conditions with the possibility to deal with potential risk factors and a fairer distribution of the principle of solidarity.

This health system must be able to symbolise an image of hope, regardless of the state of health considered.

The financial aspect is based on physiology to determine the capacity of action of an individual in a community. Evolving according to the state of health, behaviours themselves vary over time. The social component must be analysed and taken into account separately so that care is provided as needed. We distinguish three levels to be managed: the state known as being in good health, as determined physiologically by the maximum capacity of action; the state of acute illness which creates provisional dependence; and the state of chronic disease and disability, which defines a definitive physical and/or psychological dependence. In the project presented in this book, then, in my contact with the President of the Republic, I mentioned the idea of an analytical method, allowing the assessment of the capacity for action, which I called VISA (Voluntary, Incentives, Solidarity, and Anticipation). The efficiency of these four behavioural parameters varies according to levels of health and they condition individual and collective capacities to take charge. Indeed, it is necessary to anticipate, to recognise our dependence and our physical and social self-sufficiency, to be encouraged to do so and to anticipate risk situations (dependency, disabilities).

At its inception in 1946, the Social Security was well-adapted to the situation of "the post-war period" dedicated to emergency and distress. It managed acute and chronic injuries, accidents at work and their compensation, incapacities and disabilities. The means of detection were essentially clinical and there was no general ageing of the workforce and little management of dependence.

The new health system based on tripartition integrates the idea which gave rise to social security: it retains its role as manager of emergency and distress. By developing primary prevention, it facilitates the assumption of financial responsibility by the individual concerned as well as companies which conduct expensive screening. In characterising tertiary prevention, it allows the development of a social policy which takes into account the requirements created by dependence. It preserves the state's role as arbitrator, but limits its field of intervention.

In primary prevention, two poles can be distinguished: systematic prevention (without specific orientation) and prevention in the random zone. Systematic prevention must be managed by the individual, businesses and insurance. Individual effort towards primary prevention can be encouraged by a reduction in charges for health insurance. Businesses which accept this management must be eligible for a reduction in expenses in proportion to their investment. Insurance, by definition, has an interest in promoting primary prevention. The random zone is covered by health insurance according to risk.

Secondary prevention is ensured by the current system of health insurance.

Tertiary prevention relates to varying situations. We have already analysed this problem. We can add that, given the fact that a dependent person is very often attached to his/her home, environment and habits, we recommend home assistance in the case of very long-lasting chronic illness or home nursing in acute cases. Dependence certainly holds the most significant budget position and also requires long-term management. Three cases can arise: the occurrence of sudden disability interrupting a life and generating a dependency, sometimes very long term, whatever the cause; projected onset of dependence related to ageing; and the occurrence dependence associated with chronic illness. The combination of these three cases highlights the need for an equitable sharing of costs. Ageing without dependence is now a privilege, which can, of course, be due to good management of primary prevention. Unfortunately, this advantageous situation is generally not the case. It is necessary to create a mandatory dependence insurance that takes account of this. Part of this investment will be assimilated by a life insurance and transmissible to the spouse or children in the event of death. This idea implies that a part of the contributions made for retirement must be used to manage dependency. We consider that an independent structure must manage, along with social partners, a percentage of the money paid for retirement in order to guarantee the introduction of infrastructures to manage dependency. Everyone must participate according to their means towards their "dependence responsibility". Two solutions are possible: dependency managed by the hospital or in a dependency nursing home, and the dependency managed at home in the case where continued personal assistance is indispensable. The necessity of a non-medical vigilant monitoring is a major challenge. The training and remuneration of careers represents a significant budget. Because of the decision to close some hospital services, it could be wise to reclassify some hospital workers in this position. Part of this budget could be assured by retirement pensions and family benefits and also by private insurance.

Medical practice will be divided according to the three levels which were previously defined. Three categories of doctors will be jointly involved: primary prevention doctors, secondary prevention doctors and tertiary prevention doctors.

Business can play a stimulating role in primary prevention. Social Security, as it exists today, is fully maintained at level 2. It deals only with biological and therapeutic tests used in the random zone of level 1 as well as the therapeutics of level 3, leaving aside the infrastructures related to dependency. These must be managed as described previously, and it is also essential that businesses also invest in screening tests, particularly those who make large profits from the health industry, while others could obtain a reduction in some of their expenses as compensation.

We have already spoken of the importance of the awareness of the role of prevention for all the subjects, after successful discussions with the doctor, irrespective of the level of health. "To better understand ourselves" is a process that helps an individual take responsibility and reflect on life.

However, the term 'prevention' often leads to confusion. Indeed, in public health, it is usually considered that secondary prevention occurs after an event (myocardial infarction for example). However, between physiology and pathophysiology, there is continuity, although it is conventional to separate these two fields (Diagram 5). Thus, the incubation time of a disease (as in the HIV-positive of AIDS) occurs in the primary phase and the clinical expression in the secondary phase. Physiological prevention does not create a hiatus between primary and secondary prevention. Consequently, the management of a clinical pathology is integrated at its onset in secondary prevention. This has major repercussions in terms of budgeting for levels 1 and 2.

Furthermore, it is necessary to clarify the scope of medical practice. Going into the finer details, we move from a clinical vision to a genetic approach. However, it is possible to speak of prevention at all levels. There is a stratification of prevention which identifies different areas of intervention. The stage of prevention must be defined for each field of intervention (for example, physiology) to clarify the language and avoid confusion.

The medical management of innate prevention leads to an assumption of responsibility of health, thus limiting the consequences of certain risk factors likely to create physical deterioration and/or psychic acute or chronic changes. It is open to all, especially those waiting for progress in medical treatments. Prevention is a generic term which combines innate prevention and voluntary prevention. It cannot be locked into the idea of perfection. If so, it would be identified with an ideological system. Prevention is always represented in the body, even if the health condition is strongly altered. By considering that all health conditions rely on regulatory mechanisms and prevention, we do not create more opposition between one state and another.

We manage physiological balance as far as possible with our medical knowledge.

The population must get used to consulting primary prevention doctors to learn to manage their physiological needs. Education of children and parents, the management of the random zone, the existence of joint consultations with hospital practitioners, the use of specific therapies, the practice of co-prescriptions and epidemiological management make the primary prevention doctor a vital actor in the health system whose practice features the most innovative characteristics of the system.

In brief, this health system hopes to bring more clarity and insight into the management of the body, more individual and collective responsibility in response to disease, better management of the risk factors, the development of individual and collective voluntary participation and better budgeting of the system.

Founded on prevention in its noblest and most general sense, this health system could constitute real social progress. By making the different actors assume responsibility, it would establish a more effective protective system against the risk factors and would make it possible for the population to adapt their medical management. It would provide a vision of hope for everyone, whatever their current state of health.

The Ministry for Prevention

Why create a Ministry for Prevention integrating medical prevention and environmental prevention?

Prevention becomes an individual, collective, national and international preoccupation. The population has an increasing awareness of the environmental risks having an impact on health. Any voluntary and conscious threat to environmental balance could in the future become a *Casus Belli*. Even the concept of prevention is a source of balance for people, at once economic, social and political. Although it is currently neglected in India and China (against the grain of their own ancestral philosophies), it will allow a discourse in the same language with emerging nations in order to limit, as soon as possible, some risk behaviours, such as those generated by industrial production. Bringing together health and environment leads to reactive behaviour, individually and, more importantly, collectively. The creation of a Ministry for Prevention is a step in this direction by presenting a real breakthrough, both national and international.

The term 'preventive medicine' covering all medical situations removes the existing dualism between health and disease, and also between a preventive system and a healing system. It should have been in practice since Claude Bernard because of the recognition of the autonomous physiological mechanisms of regulation of our body (for example, biologists currently work on the regulation of genes). These mechanisms are active regardless of health status, but they present limited capacities for immediate adaptation. A tripartite classification of health status gives a sense of responsibility to doctors and the population and requires a change in attitudes with an epidemiologic dynamic consensus to separate the levels of care management.

Health and environmental actors would like to have a Ministry for Medical Prevention and for Environmental Prevention integrated in the Ministry for Prevention.

The Ministry for Environmental Prevention could also be divided according to a tripartite model: primary environmental prevention would correspond to the management of the ecosystem in balance with man, secondary environmental prevention to the management of serious environmental deterioration having a transitory effect on health, and tertiary environmental prevention would focus on the management of situations having a sustained destabilising effect on humans (alteration of homeostatic balance).

The alliance of a Ministry for Preventive Medicine with that of the environment within a Ministry for Prevention would clarify the management of medical and environmental risks simultaneously, bringing together and integrating health and the environment, facilitating global communication on prevention at international forums, to differentiate voluntary prevention (acquired) from independent prevention (innate), to make each individual responsible according to his state of health, specifically to deal with health in his own environment, to distinguish acute disease over time and its management on a collective level, and to deal with chronic diseases and disabilities according to a notion of predictability (levels of predictability).

Disturbing Equation

Before entering the world of pathology, medical studies teach the basic sciences. Among these, there is one that bridges the gap between the infinitely small and infinitely large, between science and philosophy of action. Through the study of body functions and expressions, physiology characterizes mechanisms of action and specific regulation. It highlights its impact on daily life, whether to explain the expression of metabolic, neuro-psychological, immune or waste functions. It therefore becomes easy to study the functional capacity of an individual through the data of physiology.

If the functional capacity of man is deeply linked to the expression of the body, it is nonetheless true that it goes beyond the body, by the instruments that we make to improve it. However, it has grown considerably since the 19th century. Man, by combining and comprehending logic systems, has always sought to increase his potential for action.

Four main areas are affected by organic interdependent technical developments: the first is the dynamic *function,* that is to say systems that facilitate movement. It has resulted in increased mobility and, consequently, in reducing distances. Bicycle, train, car, plane and have shortened the journey times, improved interchange between different countries, making the most remote areas accessible to all.

Moreover, the intellectual function has been increased by technology, and especially by the computer that has contributed to world-wide communication.

The third function of defence of the individual has been socially developed by the army to protect society.

Finally, the development of industry is always associated with waste producing pollution that has to be contained. Ecology appeared late, following waste production. Measures should be taken to reach waste disposal targets so as to maintain ecosystem balance.

The disturbing point is as follows: If we consider that tripartite social models reflect the ancient pre-industrial state of society and that social quadripartition is representative of industrialized societies, how is it the ancient Indians initiated a quadripartite society regulated by jurisdictional interactive system (Dharma)?

Conclusion

It is unusual to identify physiology with prevention. Yet physiology appears to be the result of the regulatory mechanisms of the body, and it synthesises autonomous and voluntary prevention. It defines a steady state of balance and characterises the dynamic prevention of the organism.

From nutrition and an interest in the ancient Indian social system, this book evolved into the elaboration of a health system project founded on prevention. The central theme of the book revolves around individual and collective control physiological risk, whether that involves the resolution of being overweight, the management of various states of health, or that of a society generating high risk.

On a medical level, the management of health conditions reveals three levels of prevention. Primary prevention manages the functional balance of the body and should integrate therapeutics to fight against risk factors. This is currently not the case. Secondary prevention deals with acute disease throughout its duration. Here, the role of the Social Security comes to prominence. As the treatment is limited in time, it can be easily budgeted. Finally, tertiary prevention must take charge of chronic diseases and disabilities according to a notion of predictability. There are differences between a crippling accident of youth and the disability of old age, of prediction and of the capacity for care! By calling on voluntary participation, incentives, solidarity and anticipation, as well as by determining the dependence, it is possible to design a health system based on a physiological definition of prevention. The Ministry for Medical Prevention would be composed of three authorities in charge of health, acute diseases and emergencies, and chronic diseases and disabilities.

The human body can be represented by a circle divided into four equal quadrants, each having a specific physiological function, and the ensemble constitutes a steady state of auto-regulation. To increase our capacity of action we have created extra-organic prostheses (EOPs) for each function. These prostheses represent risk before, during and after each stage of production and utilisation. Human organisation has formed social group around these functions and their prostheses. These include those involved in the production process, in human activities, in ecology, in the army, in philosophy, religions, etc. The social change that results is that of a multidisciplinary organisation integrating all parties directly concerned with the existence of each of these prostheses. Yet, even though these prostheses bring benefits, they also generate risks which have undeniable social repercussions.

The evolution of our society highlights four major risks linked with our functions and the EOPs that we use. The increase in industrial production has generated physical, chemical and biological wastes. Their management requires a specialisation, which has given birth to a group of individuals devoting their professional lives to the research of waste treatment techniques in order to maintain environmental balance. This fourth social group completes the groups administering the metabolic, neuropsychological and immunological prostheses.

It is in studying the stability of the Indian social system that I became aware of the importance the ancient Indians gave to physiological functions. Indeed, the functional division of Indian society has remained invariable in spite of internal secular drift and the thousands of years which have elapsed since their creation.

Based on the principle that a quadripartite social system defines a society at high risk, I was interested to seek the reasons which caused the establishment of such a system and its survival up to now. Today, we are faced with face the following situation: the people of India have passed on to us, through a social system which has evolved over centuries into the castes, a revelation about an event which would have been taken place in ancient times.

It should simply be noted that it would have been caused by man and that we have lost any direct trace of it. However, thanks to the rigour of oral transmission, it has reached us in the form of behavioural advice, borne by the Indian people as a burden over the centuries. The original objective of this system would thus have been to protect us, to warn us of a risk of disaster caused by man. It is through what we call the antithetical memory that we seek access to the event.

Survivors or initiators of this very distant event (some would use the term 'mythical' but I am opposed to it because it would remove the substance of the message) have sought a type of organisation likely to maintain social balance in response to a pre-existing social disorder. Yet everything happened as though they knew (how?) that man is a self-regulating system for which it is possible to apply certain rules of cybernetic modelling. They knew also that physiological adaptation has its limits, and that to go beyond those, we could regress or even disappear. All these parameters seem to have been integrated by theorists of the system. Imagining that they modelled the body in four major functions, they created a system to facilitate it organisation. We hypothesised that each function has its EOP around which a man would organise himself. According to the theory in the first chapter of the book, the establishment of a quadripartite social system corresponds to the division of a society with high risks. However, ancient Indian society did not correspond to our definition of a high-risk society. It can be identified as a low-risk society according to the Dumézil model[1]. This odd choice appears all the more theoretical because it changed over the centuries to become a tripartite society, so dear to George Dumezil in his studies on Indian mythology.

There was originally a stated desire to establish a theoretical system that was not suited to the established society. What was the purpose, or what was the ulterior motive? It seems that fear of seeing the same errors occur as those which had caused the disaster led these theorists to anticipate all situations, even those not yet existing at the time, knowing intuitively that one day this system would

1. Georges Dumézil, *Mythes et dieux des indo-européens* (Myths and Gods of the Indo-European), Flammarion, 1992, p. 95.

appear to be completely suitable. There was a prosthetic inertia which they were aware of. Yet, today, the four EOPs function intensely. The development of waste disposal, coupled with the metabolic sector, justifies the existence of a group of useful organisers, recognised and respected as in the Vedic period.

It is necessary to speculate in order to try to understand the reasons driving the theorists to adopt a system which could not fully function in a quadripartite society. It can only be a reaction of fear whose justification is at the level of the message transmitted.

"For story to appear in places where it is unknown, it takes a jolt of high amplitude which contains within it the conditions of its comprehension and which spreads in wide, concentric waves across the centuries."[2]

What is certain is that only high-risk society is capable of modifying homeostasis to the point of creating the conditions for the elimination of mankind. It is a dangerous society which carries within it the seeds of its own demise.

It seems completely probable to me that the ancients, according to an oral tradition, had knowledge of human behaviours having endangered humanity. Their long-term vision of the future appeared to be the only remedy to prevent a similar event happening again. They must have considered that the system they advocated would not be altered by time. It was a question of establishing a long-term project based on unquestionable and lasting data.

Some points raise a series of questions:

- The first point that surprises me is the structure of reasoning, which draws on physiological data that we

---

2. *The Mahabharata* 1, op. cit., p. 30.

have recently acquired, constituting the framework of a social system whose content has certainly drifted over the centuries, but whose foundations remained the same. It is all the more astonishing that these initiators seem to have made a connection between a physiological system, prosthetic organisation and social risks. In other words, the accuracy of their assessment of these social risks. Moreover, the Indians stress the universality of their reasoning following the example of physiology which applies timeless laws. This decision was accompanied by messages of wisdom and an inexplicable feeling of underlying guilt which, to be understood, we must refer to in past texts.

- The second, more intriguing point is about the evolution of the Indian social system over centuries: the system would weaken if it did not reject the social group least representative, the Sudras, who have now become, in our modern societies, a very highly representative group. It is the caste of the Sudras which deals with waste disposal. However, the fundamental normative structure of the system persisted and this attitude suggests that the experience of the past and the fear of the future would have prevailed on the establishment of a tripartite system. Is this an inertial reaction taking into account the human capacity to develop quadripartite systems dangerous to himself/herself? This obsessive fear seems to have been sufficiently powerful to provide a basis for a theoretical project which has endured. This theoretical provocation marked the memory of a former event having been responsible for a human disaster.
- The third disconcerting point is the intense memorising effort required by the Brahmans and

accepted by the people of India over the centuries, repeated to each generation that the revelation to be conveyed was of such gravity that the message should not be distorted. Thus, it was necessary to learn the texts by heart, even if it meant that not all was understood.

- The fourth point, and not the least, is content of those messages which generate an inexplicable distrust of all men, especially those not achieving Dharma, that is to say not respecting the rules.

Finally, it is completely surprising that the Indian people, like the Chinese people moreover, have expressed a real obsession with balance, as though they sensed that the world was only on borrowed time.

While the Indians developed a social system corresponding to a balanced management of physiological compartments of the body, the Chinese built their civilisation on an atomic design (Einstein) founded on the respect of yin-yang balance highlighting the complementary nature of matter and energy.

Is it possible to think that these two lines of reasoning stem from the same root cause? According to our Western education, it is possible to interpret the Indian message dealing with the management of balance as being founded on respect for the physiological homeostasis of the body. In the same epoch, the Chinese preached the balancing of energy and matter, between Yin and Yang. This objective reveals a level of knowledge of physiology as in physics completely up-to-date.

It is also particularly surprising to note that a few centuries ago the level of scientific knowledge in India, as in China, was much higher than ours, while their technical implementation did not progress. A plausible explanation to understand the absence of the material evolution of these civilisations, unlike the West, seems to stem from the field of philosophy which was adopted. Would it be this famous event which generated it? How can we not be astonished by the many Chinese discoveries and inventions listed in the encyclopaedia of Joseph Needham[3] and which have never been used?

We are now at a turning point in history in a very paradoxical economic situation: on one side the West is beginning to realise that we are endangering the planet; on the other, the Indians and the Chinese, who have transmitted a message of behavioural caution and of balance, have put aside their worries while trying to take advantage of materialism and its apparent benefits. They would like to forget their philosophy and seem to think it has slowed their expansion for centuries. However, the message remains clear and it is the West, in its materialistic maturity and environmental awareness, which reacts. What are we doing? Where are we going? What will happen to the generations following this materialistic turmoil?

Should we conclude that this reflection on physiology can lead to wisdom? If we look at the evolution of our societies, we realise the power we give EOPs. Overdevelopment or the underdevelopment of these prostheses is a source of social instability of which we do not measure all the effects, but which we see occurring with concern. What remedy do we suggest? Of course, there are conferences and some corrections, but we do not fundamentally change social and political behaviour in risk

3. Joseph Needham (dir.), *Science and Civilisation in China*, Cambridge University Press, 7 volumes, started in 1954.

management. At the heart of the problem there is the human being who obeys physiological laws presenting unknown limits of adaptation. Human physiological balance is deteriorating due to environmental changes. It is necessary to talk about the environment in the broad sense. That is to say, for example, in the sense where we believe that in vitro fertilisation constitutes a temporary modification of the environment. Is it necessary, then, to worry about medical progress? The future will give us the answer, but I fear we are realising too late, that we have sometimes been irresponsible. If we assume that we follow a law of evolution straight upward, we can only progress. However, if there is a regulation of genes which does not reflect our will, but the environment, could it be possible that we start a phenotypical regression? If genes determine, in their "wisdom", not to express themselves as before, what will happen to man?

Is there not a lesson to extract from Indian history?

Is our philosophy a barrier to progress and modernisation?

I would say no, because the vast majority of Indians are realistic, and far from being opposed to change ... India remains deeply rooted in its past, but it is also receptive to the latest discoveries which have revolutionised life. It draws on two streams without losing balance: India has always been an enigma."[4]

Madeleine Biardeau, former Director of Advanced Studies in Ecole Pratique des Hautes Etudes, seems to reply to this question in echo: "India would thus have in its oldest and most authoritative tradition, all which is necessary to renew

4. Indira Gandhi, Jean-Louis Nou, op. cit., p. 9.

its vision of the world and to build an ideology capable of spiritually nourishing a 'modern' society."[5]

Perhaps their message will be heard and we will use their vision of society to protect us from risks which we run. But will we have the social and political courage to face the world we will leave? If this hypothesis proves true, it would be interesting to try to better understand the meaning of life. Solutions are emerging but the concept of risk needs to be not only studied but managed harmoniously.

The ancient Indian social model seems to come close to a vision foreseen by Claude Bernard. The discovery of the regulatory mechanisms of the body, highlighted by modern biology, leads us to question the study of our needs and functions. The management of EOPs, in terms of their design, use and the eliminations which they cause directly and indirectly, pertains to a world inhabited by experts specialising in their management. Today's policies must ensure a share of responsibility in managing these risks in order to maintain human physiological balance. The adoption of a social quadripartition and interactive logistics would favour the management of a high-risk society. The democratic society of the future would require a distribution of responsibilities and inter-group jurisdiction. We may be on the verge of a profound political change that could prevent a disaster caused by human activity. Forget the secular drift of the ancient Indian model in order to understand its relevance in a high-risk society.

Undoubtedly, the balance of man in his environment seems a prerequisite for finding the capacity to formulate the crucial questions and, possibly, to be able to respond. Does progress not depend on the evolution of questions?

---

5. Madeleine Biardeau, op.cit., p. 262.

"Philosophy, by unceasingly agitating the inexhaustible mass of unsolved questions, stimulates and maintains this healthy movement in science."[6]

There is therefore an interaction between the theories of evolution, human-environmental balance and the questions which we are led to ask, the answers remain in the field where the reflection of each person is expressed according to his/her knowledge and by his/her choices.

---

6. Claude Bernard, op. cit., p. 308.

Appendix

Presentation of the project of a health system based on prevention with Mrs Marie-Claire Carrère-Gée, Social Adviser to the President of the Republic, within the framework of the Elysée in October 2002.

This project carries with it a sense of evolution and not a sense of revolution. It traces ways to reorganise the health system.

System Philosophy

- Man obeys the laws of internal regulation (physiology) and external control (environment).
- Our morphological identity is the result of a dynamic state of equilibrium (homeostasis).
- Our behaviours are consistent with our state of health.
- Our capacities of action are functions of our state of health.

Sequence of Medical Reasoning

- The human body has an innate system of prevention.
- Human health depends on the management of its balance.
- Doctors must ensure health.

Definition of Medical Prevention

- Dualistic definition: set of measures taken in order to avoid disease.
- Epidemiologic definition (OMS): study of the conduits to be held according to three situations: primary, secondary and tertiary.
- Physiological definition (upstream): management of all of the regulatory mechanisms allowing the maintenance of a balanced health condition (steady state).

To differentiate classification of prevention in public health from that of the prevention in physiology:

- Public health prevention: hiatus between primary and secondary prevention, management of secondary effects and chronic consequences of a disease
- Physiological prevention: centred on physiological balance, consider the non-event, no hiatus between primary and secondary prevention.

Modelled representation of variations of the health condition: to note the axis of time, clinical axis SH1, SH2, SH3, corresponding control loops, therapeutic 1,2,3, the absence of clinical signs, incubation or balance, estimated dependence (SH1) with paraclinic signs (+ or –), acute clinical signs and paraclinic signs, acute imbalance, provisional dependence (SH2), chronic clinical signs and paraclinic signs, chronic imbalance, final dependence (SH3).

Metaphorical representation of the human organism and its physiological functions in an environment (quadripartition of a circle).

Why modify the health system?

- For better coherence
- For better risk management
- For better social management
- For better financial management

Objectives of the Health System

- Medical
- — To differentiate and treat medical acts on a hierarchical basis;
- — To define the sphere of activity of primary prevention and its separation from secondary prevention;
- — To determine the field of intervention of secondary prevention to an action limited in time;
- — To specify the field of intervention of tertiary prevention
- — To choose a classification of prevention accepted by all; and
- — To show the medical advantages of a tripartition of the health system
- Social
- — To clarify information; and
- — To modify behaviours about disease: method VISA.
- Economic and Financial
- — To bind the physiology of behaviours to the budgeting of the system: method VISA;
- — To use a tripartition of the system to rebalance the budget;
- — To delimit the role of the actors according to each level; and
- — To make the government the arbitrator of the system.

Behavioural Methodology of Budgeting

- Four parameters VISA: Voluntary Participation, Incentive, Solidarity, Anticipation
- Combinatorial analyses
- Shared responsibilities
- A fair distribution of budgets

Expected Results

- To homogenise the entire health system
- To make all actors take responsibility
- To balance the budgeting

Installation of the Health System: Three States of Health

- Modelled representation of the variations of analyses health conditions.
- Field of definition of the primary health condition (SH1)
- Field of definition of the secondary health condition (SH2)
- Field of definition of the tertiary health condition (SH3)

Primary Prevention Policy

- The infant: the paediatrician
- The teenager: the annual family notebook of health prevention
- The adult
- Actors of primary prevention: the individual concerned, primary prevention doctors, the business, mutual insurance companies, insurers – family contract of health prevention

Policy of Business in Primary Prevention

- Implication in primary prevention
- No-claims bonus in secondary prevention (disease)
- Secondary prevention policy
- Management of the treatments for acute diseases
- Duration and limited budgeting

Tertiary Prevention Policy

- Treatments of chronic diseases and disabilities
- Management of dependence
- Make responsible: forecasting dependence and capitalisation of retirement health items

Three Actors of Health

- Primary prevention doctor
- Secondary prevention doctor
- Tertiary prevention doctor

How to define a primary prevention doctor?

- Stage 1: consultation, a clinical study
- Stage 2: study of physiological balance and the environment
- Stage 3: an epidemiologic management of the risks
- Stage 4: therapeutic range and a curative practice often unnoticed

Therapeutic Approach in Primary Prevention: An Evolution

- The therapeutic arsenal
- The question of de-reimbursement
- A redistribution

The Implications of the New Approach

- Primary prevention and health professions
- The "new general practitioner" must also be a preventer
- A new trade about health by the environment
- Primary prevention and the society of risk
- An increasing calling into question of "non-prevention" public
- A factor of economy for the social security
- Prevention and organisation of care
- An integrated primary prevention system/general care
- An integrated primary prevention system/hospital
- An integrated primary prevention system/environment

Conclusion

- A complete and homogeneous system of risk management
- Three insurances (prevention, care, dependence)

Glossary*

Agni : He is the god of fire, the principle male, appears in the kuṇḍa, the home of the altar, the image of the female sex. The woman is the guardian of the home, the priestess of Agni.

Asuras : Although regarded as demons, the asuras have the distinction of being the elder brothers of the gods. Meaning "divine", they represent an ancient Vedic term the supreme spirit. Thus, Varuna, Mitra, Indra and Agni, identified with the supreme gods, are

* This glossary has been achieved by drawing on the writings of Madeleine Biardeau, former director of advanced studies at the Ecole Pratique des Hautes Etudes, in particular thanks to the book entitled *Hinduism,* Flammarion, 1995, of Jean-Claude Carrière in the *Dictionary in Love with India,* Plon, 2001, the collaborators of the Asian collection, myths and traditions led by André Akoun, Brepols, 1991, by Alain Daniélou in *Myths and Gods of India*, Flammarion, 1992.

qualified as asuras. There is no logical explanation for this transformation of good gods into bad, unless it is that they are said to have usurped the place of the gods in heaven or incarnated themselves as princes on earth to oppress Brahmans. So, they are the personification of adharma and an avatar of Vishnu should appear to rid the world of them.

Avatara : Descent of God in our world each time the socio-cosmic order deteriorated (adharma) to restore order, to restore the dharma and to save man. He appears at the end of cosmic cycles infinitely renewed. Brahma, Vishnu and Shiva are the representatives at the beginning, during and at the end of a cycle.

Bhakti : Path of liberation of the infernal cycle of rebirth which is identified with the sacrifices of sannyasis, concerning not only Brahmans but all individuals who want to enter into a privileged relationship with God. She makes it possible to obtain deliverance.

Bhima : The second of the five Pandavas in the Mahabharata. He is the son of vedic god Vayu (the wind).

Bhisma : Paternal grand-uncle of Pandavas and of their enemy cousins (Kauravas) in the Mahabharata. He gave up marriage and the throne to allow the king, his father, to marry the woman he desired and to assure royalty for his children.

Brahman : "At the lower level, the neutral term means the ritual formula, then the Vedic knowledge, the science and the power of the Brahman, the Veda itself or its symbol, the syllable om. Male, it is the name of one of the priests of the Vedic sacrifice. At higher levels, it is the

Absolute (it is distinguished by the use of a capital letter). This Absolute (neutral) gives a divine personage, whom we designate here by the nominative form masc.: Brahmâ."[1]

Deva : "God", particularly the gods of Vedic sacrifice.

Dharma : Socio-cosmic order on which the existence of the universe depends. The disorder, adharma, brings an end to the world. Maintaining the Dharma is more particularly entrusted to the Brahmans. We believe that the Dharma is a jurisdiction between the four Varnas which maintains the physiological balance of the human body or its homeostasis, since it is related to the environment in its broadest sense, which is with the cosmos.

Kali : It is an almost childish terror which we cannot manage to comprehend. "Kali is the true fear because we do not know who it is"[2]. The word Kali means "The Black One" a merciless image of time which leads us all to death. One of the terrible names of the virgin Goddess.

Kaliyuga : The fourth and last yuga, the worst, one in which we live. It is the "iron age" of Hindus.

Kalpa : A day and a night of Brahmâ constitutes a kalpa: each one of these kalpa are subdivided in thousand quaternary cycles and each one of these cycles into four yugas. Man currently lives in the fourth yuga, before last hour before the end of a cycle which will put an end to the current world.

---

1. Madeleine Biardeau, op. cit.
2. Jean-Claude Carrière, *Dictionary in Love with India,* Plon, 2001, p. 193.

Karma : It would be towards the fifth-sixth century before our era that the concept of retribution for actions (karma) would have appeared whereas the belief in transmigration continues. If any act is done for a future result, the karma is also what condemns to be reborn to collect the "fruits" which did not "mature" in the present life.

Krishna : The ninth avatara of Vishnu. The era of Kaliyuga begins the day he was mortally wounded by a hunter who mistook him for a stag. In the Mahabharata, he appears mostly in the guise of a warrior (Ksatriyas) devoid of royal power, but very influential advisor of the Pandavas. He is indeed an unpredictable being, defending life, an elusive warrior.

Ksatriya : Second varna of the Brahman society, that of warriors and princes.

Mahabharata : This is a voluminous epic poetic work which was at first transmitted by oral tradition before being written in Sanskrit shortly before our era by Brahmans. We do not know the moment when this fabulous story occurred which was enriched by new texts over the centuries. If some think it is a collective work, others attribute it to one person because of the coherence of the whole. "It seems every moment to recognise the brilliant presence of a persistent author, aware, responsible and supremely glorious in anonymity."[3] This work is told by a disciple of the legendary Vyasa who mingles with the story by making love with two princesses who will give birth to two kings whose

3. Jean-Claude Carrière, op.cit., p. 267.

children will make war to the point of jeopardising humanity.

Purana : The first Puranas, which were texts belonging to the tradition, were worked out in the drafting of the Mahabharata. Theoretically, they recount the origins of humanity and of Indian history. In fact, they contain all that a Hindu must know to act correctly in all circumstances.

Rudra : "Lord of Tears", he is the fire which burns and personifies the sun. He is a terrible force of nature who displays a duality between his destructive side and his beneficial aspect. This is the most common name of the fiery form of Shiva in the Vedic literature.

Sannyâsin : This state of life is that of "giving up", and more particularly of the Brahman who renounced the ritual and his function in the world to become a monk. It is a state of life (ashrama) where asceticism can free them from the infernal cycle of rebirths. If the ascetic does not belong to the varna Brahman, he is called sâdhu and he is often dreaded because his magical powers are feared.

Shiva : One of the two great gods of the bhakti. His role as catalyst of impurity in vedic sacrifice (under the name of Rudra) caused him to be linked more to forms of renouncement which diverge the furthest from orthodoxy.

Sûdra : Servants who are not allowed in the Vedic ritual and who constitute the fourth class (varna) of Brahman society.

Vaishyas : Third varna of the Brahman society, farmers and retailers.

Varna : "Class." The Brahman society ideally is divided into four varnas. The term indicates at the same time a social function, a statute and a religion specific to the vedic ritual.

Veda : Litt. "knowledge". It is knowledge in the greatest degree of excellence, the Revelation. The texts which constitute it are divided into four major groups which represent the four Vedas: Rigveda, Yajurveda, Atharvaveda and Samaveda. The first three correspond respectively to three classes of priests of Vedic sacrifice. The fourth, which contains many magic formulas, is regarded as inferior.

Vishnu : One of the two great gods of the bhakti making part of the Hindu trinity. The word comes from the root "vish" meaning "to penetrate, to be everywhere". In the old Vedic culture, Vishnu is the omnipresent one, the immanent one, he who has the knowledge of the cosmic laws. He can unite in power. He represents the power by which all the things exist. While Shiva presents a destroying principle, Vishnu expresses the continuity of the universe. Vishnu, by the faith and knowledge which he transmits gives rise to a religion.

Yuga : Name of cosmic periods grouped by four, each indicating a given state of perfection or degradation of the dharma. The avatara of Vishnu generally appears in the passing of one yuga to another.

Bibliography

Akoun, André (dir.), *L'Asie, Mythes et Traditions,* (Asia, Mythes and Traditions), Brepols, 1991.

Amrani, N., F. Commandre, E. Fonaris, N. Makhtar, P. Vanuxem, *Les problèmes de thermorégulation chez les sportifs (The Problems of Thermoregulation in Athletes),* Fac. Médecine lab. Physiologie respiratoire, Marseille 13385, France, Médecine d'Afrique Noire, 1990, Vol. 37, NE11.

Apfeldorfer, Gérard (under the direction of), *Traité de l'alimentation et du corps* (*Treaty of Food and the Body*), Editions Flammarion, Paris, 1994

Aristotle, *La Métaphysique*, Agora, Les Classiques, Presses Pocket, 1991 (*Aristotle, Metaphysics*, Books Gamma, Delta and Epsilon, Second Edition, Translated with notes by Christopher Kirwan, Nov. 1993).

Baudouin, Bernard, Le *Védisme, l'éveil de la spiritualité indienne,* (*The Vedas: The Awakening of Indian Spirituality),* Editions de Vecchi S.A. , 1997.

Bateson, Gregory, Rodney E.Donaldson, *A Sacred Unity : Further Steps to an Ecology of Mind*. Harper Collins, 1991 (*Une unité sacrée, Quelques pas de plus vers une écologie de l'esprit*, Editions du Seuil, 1996).

Bateson, Gregory/Ruesch Jürgen, *Communication: The Social Matrix of Psychiatry*. New York: W.W. Norton & Company, 1951 (*Communication et société*, Editions du Seuil, 1988).

Baillet, Jack/ Nortier, Erik, *Précis de physiologie humaine* (Handbook of Human Physiology), Ellipses, Edition Marketing, 1992

Beck, Ulrich, *Risikogesellschaft. Auf demWeg in eine andere Moderne Frankfurt am Main, Suhrkamp, 1986 (La société du risque. Sur la voie d'une autre modernité*, Paris, Editions Flammarion, 2001) (English Edition: *Risk Society: Towards a New Modernity*, London, Sage, 1992).

Bernard, Claude, *Introduction à l'étude de la medecine expérimentale*, Champs Flammarion, 1984 (*An Introduction to the Study of Experimental Medicine*,1865. First English translation by Henry Copley Greene, published by Macmillan Co., Ltd., 1927; reprinted in 1949).

Biardeau, Madeleine, *L'Hindouisme, Anthropologie d'une civilisation*, Flammarion, 1995. (*Hinduism: The Anthropology of a Civilization*, OUP India, 1990).

Bigé, Luc, *L'homme réunifié* (Man Reunified), Editions du Rocher, 1995.

Canguilhem, Georges, *Le normal et le pathologique*, Quadrige PUF, 8ème Edition, 1999 (*The Normal and the Pathological*, Zone Books, 1991).

Cannon, Walter, *Wisdom of the Body*, W.W. Norton and Company, Inc.. Rev. and Enl. Edition, 1963 (*La sagesse du corps*, Paris, Editions de la Nouvelle Revue Critique, 1946).

Carrière, Jean-Claude, *Dictionnaire amoureux de l'Inde* (*Dictionary in Love with India*), Editions Plon, 2001.

Capra, Fritjof, *The Tao of Physics*, Shambhala Publications, 1975; Shambhala, 4th edition 2000 (*Le Tao de la physique*, Editions Sand 1985, Tchou, janvier 2006).

Cassé, M. *Du vide et de la création* (*The Void and of the Creation*), Editions Odile Jacob, 1993.

Confucius, *Entretiens avec ses disciples*, Editions Flammarion, 1994 ; (The Analects of Confucius : Or, the Conversations of Confucius with His Disciples and Certain Others (The World's Classics), Oxford University Press, 1951).

Corvol, Pierre, Nicolas Postel-Vinay, *Le retour du Dr Knock* (*The Return of Dr Knock*), Editions Odile Jacob, 2000.

Crozier, Michel, *La société bloquée* (*The Company Blocked*), Editions du Seuil, 1970.

Damasio, Antonio R., *Le sentiment même de soi : corps, émotions, conscience*, Odile Jacob, Paris, 1999 (*The Feeling of What Happens: Body and Emotion in the Making of Consciousness*, Mariner Books, First Edition, 2000).

Danielou, A., *Mythes et dieux de l'Inde, Le polythéisme hindou*, Editions Flammmarion, 1992 (*Gods of India: Hindu Polytheism*, Bollingen Fondation, Princeton University Press, 1964. Inner Tradition International, New York, 1985. New edition without sanskrit texts under the title *The Gods of India*, 1986).

Debray-Ritzen, Pierre, *Claude Bernard ou un nouvel état de l'humaine raison* (*Claude Bernard or A New State of Human Reason*), Editions Albin Michel, S.A., 1992.

Drouin, Jean Marc, *L'écologie et son histoire : Réinventer la nature* (*Ecology and its History*), Flammarion, 1993.

Dumézil, Georges, *Mythes et dieux des indo-européens* (*Myths and Gods of the Indo-European*), Editions Flammarion, 1992.

Dumont, Louis, *Essais sur l'individualisme, Une perspective anthropologique sur l'idéologie moderne*, Editions du Seuil, Points Essais, 1991 (*Essays on Individualism: Modern Ideology in Anthropological Perspective.* Chicago: University of Chicago Press.,1986).

Eliade, Mircea/ Couliano, Ioan P., *Dictionnaire des religions (Dictionnary of Religions)*, Plon, 1990.

Eliade, Mircea, *Aspects du mythe*, Gallimard, 1963 (*Myth and Reality* (trans. Willard R. Trask), Harper & Row, New York, 1963).

Eliade, Mircea, *Histoire des croyances et des idées religieuses, de l'âge de pierre aux mystères d'Eleusis*, Payot Editions, 1976 (*A History of Religious Ideas*, Vol. 1 (trans. Willard R. Trask), University of Chicago Press, Chicago, 1978).

Eliade, Mircea, *Le mythe de l'éternel retour, archétypes et répétition*, 1949. (*The Myth of the Eternal Return: Cosmos and History* (trans. Willard R. Trask), Princeton University Press, Princeton, 1971).

Emerit, J.E., *L'acupuncture traditionnelle* (*Traditional Acupuncture*), Guy Trédaniel, Editions de Maisnie, 1986.

Etienne, Jacqueline, *Biochimie, génétique, biologie moléculaire (Biochemistry, Genetics and Molecular Biology)*, 5ème Edition, Masson Paris, 1987, 1999.

Filliozat Jean (en collab. avec L. Renou) *L'Inde Classique. Manuel des études indiennes*, T1, Paris, Payot 1949; T2, Paris, EFEO, 1953.

Foucault, Michel, *La naissance de la clinique*, Paris, Presse universitaire de France, 1962 (*The Birth of the Clinic*, translated by A. Sheridan Smith, New York: Panthéon, 1973).

Gandhi, Indira/Nou, Jean-Louis, *Inde, Hommes Rites et Dieux (India, Man, Rites and Gods)*, Edita S.A. Lausanne (Suisse), 1978.

Garnier-Delamare, *Dictionnaire des termes de médecine, (Dictionary of Medical Terms)*, 22ème édition rev. et augm. (18 April 1996)

Gould, Stephen Jay, *Dinosaurs in a Haystack*, New York: Harmony Books, 1995 (*La foire aux dinosaures, réflexions sur l'histoire naturelle*, Editions Le Seuil, 1993).

Gould, Stephen Jay, *The Panda's Thumb*, New York: W.W. Norton, 1980 (*Le pouce du panda*, Edition Grasset et Fasquelle, 1982).

Grenier, Jean, *L'esprit du Tao* (*The Spirit of Tao*), Editions Flammarion, 1973.

Grmek, Mirko D., *Le legs de Claude Bernard (The Heritage of Claude Bernard)*, Editions Fayard, 1997.

Hagège, Claude, *Halte à la mort des langues*, Editions Odile Jacob, novembre 2000, (*On the Death and Life of Languages*, translated by Jody Gladding, Yale University Press, 2011)

Jenicek, M. / Cleroux, R., *Epidémiologie, Principes, Technique, Applications* (Epidemiology, Principles, Technology and Applications), Edisem, 6ème édition, Editions Maloine S.A.

Jullien, François, *Figures de l'immanence, Pour une lecture philosophique du Yi King*, Le classique du changement (Figures of Immanence, Philosophical Reading of the I-Ching, the Classic of Change), Editions Grasset, 1993.

Le grand guide de l'Inde (The Great Leader of India), Editions Gallimard, 1998.

Lazorthes, G., *Sciences humaines et sociales – L'homme, la société et la médecine (Social Science – Man, Society and Medicine)*, 5ème edition révisée et étendue, Editions Masson, 1993,1998.

Lovelock, J.E., *The Ages of Gaïa: A Biography of Our Living Earth*. New York: Norton, 1995 (*La terre est un être vivant : L'hypothèse Gaïa*, Editions Flammarion, 1993).

Le Mahabharata – T1, Livre I à V (*The Mahabharata*, Books I to V), translated by Madeleine Biardeau, Jean-Michel Peterfalvi, Editions Flammarion, 1985.

Le Mahabharata – T2, Livre VI à XVIII, (The Mahabharata, Books VI to XVIII), translated by Madeleine Biardeau, Jean-Michel Peterfalvi, Editions Flammarion, 1986.

Le Manuel diagnostique et statistique des troubles mentaux, (*Diagnostic and Statistical Manual of Mental Disorders,* American Psychiatric Association, DSM-IV), Paris, Editions Masson, 1996.

Morin, E./Piatelli-Palmarini, M., *L'unité de l'homme – Pour une anthropologie fondamentale (The Unity of Man – A Basic Anthropology)*, Editions Le Seuil (Paris), 1974.

Peretti – Watel, Patrick, *Sociologie du risque* (*The Sociology of Risk*), Editions Armand Colin, Paris, 2000.

Pasteur, Louis, *Ecrits scientifiques et médicaux (Scientific and Medical Writings)*, Editions Flammarion,1994.

Reeves, Hubert, *L'heure de s'enivrer, L'univers a-t-il un sens?* (Time to Get Drunk, Does the Universe Make Sense?) Editions Le Seuil, 1986.

Renou, Louis, *L'Inde classique: manuel des études indiennes* (collab. Jean Filliozat), Paris, Payot 1947 [i.e., 1949] – 53.

Rey, Alain, *Dictionnaire historique de la langue française, (Historical Dictionary of French Language)*, 1992

Ricoeur, Paul, *La mémoire, l'histoire, l'oubli (Memory, History and Forgetfulness)*, Editions Le Seuil, 2000.

Rinpoché, Sogyal, *The Tibetan Book of Living and Dying,* HarperOne, 1994 (*Le livre tibétain de la vie et de la mort,* Editions de la Table Ronde, Paris, 1993)

Sciences et Vie, *comment notre corps maintient ses équilibres vitaux (Sciences and Life, How Our Body Maintains its Vital Balance)*, par D.Diaz, nE361, octobre 1947.

Seignalet, Jean, *L'alimentation ou la troisième médecine* (Food or the Third Medicine), François-Xavier de Guibert, 1996.

Sgreccia, Elio, *Manuel de bioéthique, Les fondements de l'éthique biomédicale (Bioethics Manuel, The Foundation of Biomedical Ethics)*, Editions Mame-Edifa , 2004.

Shri Aurobindo, *The Bhagavad Gita or The Song of Gods* (translation and commentaries), Lotus Press, 1996, (*La*

Bhaghavad Gita, (traduction et commentaires), Editions Albin Michel, 1970).

Sicard, Didier (coordination), *Travaux du Comité Consultatif National d'Ethique (Proceeding of the National Advisory Committee of Ethics)*, Presse universitaire de France, 2003.

Spinoza, Baruch, *Œuvres III, Ethique*, GF, Editions Flammarion, 1965 (*Ethics*, in Edwin Curley, translator, The Collected Writings of Spinoza (Princeton: Princeton University Press, 1985), Vol. 1).

Temple, Robert K.G., *China: Land of Discovery and Invention*, Patrick Stephens Ltd, 1986 (*Quand la Chine nous précédait,* Présentation de Joseph Needham, 3000 ans de découverts et d'inventions chinoises, Bordas, 1987).

Thom, René, *Prédire n'est pas expliquer (To Predict Is Not to Explain)*, Editions Flammarion, 1989.

Varenne, Jean, *Mythes et Légendes*, extraits des Brâhmanas, Paris, Gallimard, 1968.

Watzlawick, Paul, La *réalité de la réalité: confusion, désinformation,communication,* Editions Le Seuil, 1978 (*The Invented Reality: How do we know What we Believe we Know? Confusion, Misinformation, Communication*, New York: W.W. Norton & Co., 1984).

Wachthausen, Jean-Luc: *'James Cameron, le visionnaire du cinéma' ('James Cameron, The Visionnary of Cinema'*), Le Figaro, 26 Decembre 2008, p.21.

Wills, Christopher, *The Wisdom of the Genes: New Pathways in Evolution*, Basic Books, 1991 (*La sagesse des gènes, Nouvelles perspectives de l'évolution*, Editions Flammarion, 1991).

Index